C0-AOC-366

BEACH PATTERNS

The World of Sea and Sand

BEACH PATTERNS

The World of Sea and Sand

by STELLA SNEAD

Introduction by Gyorgy Kepes, Director, Massachusetts Institute of Technology Center for Advanced Visual Studies

Barre Publishing
Barre, Massachusetts 1975
Distributed by Crown Publishers, Inc., New York

Inquiries should be addressed to Clarkson N. Potter, Inc.,
419 Park Avenue South, New York, N.Y. 10016.
Published simultaneously in Canada by General Publishing Company Limited
First edition
Printed in the United States of America

Designed by HELGA MAASS

Library of Congress Cataloging in Publication Data

Snead, Stella.
Beach patterns.

1. Beaches—Pictorial works. 2. Marine
biology—Pictorial works. I. Title.
GB451.S58 551.4'5 75-19384
ISBN 0-517-52403-1
ISBN 0-517-52404-X pbk.

Photographic Note

The photographs in this book were taken over a number of years, say from about 1960, and with several different cameras, but always 35 mm single lens reflex, and usually with the aid of a light meter either on the camera or separate. At first I had a Kodak Retina Reflex, then a Contaflex. With both of these I used what they provided for closeup work—a screw-on-to-the-lens attachment which looked like a filter. It was not a perfect instrument. Then I came to own a Nikon F and later an FTn, two in fact, and with them goes the most exquisite of lenses, the Micro-Nikkor f/3.5. 55 mm. With this lens anything can be done from infinity to a few inches and with the extension tube provided, even closer. But when using the tube one needs a tripod as well as perfect peace. Sometimes I used a tripod anyway, to be sure of exact precision. With many of my really sharp negatives I have been able to make 20″ × 30″ prints.

As is said in the text, one wanders a beach and looks down. When possible it is usually best to take the pattern or object, when small enough, with the camera held directly above, but of course long shots are in order too. The selection of a single pattern often makes a more successful photograph than a conglomeration of designs. When the pattern is too large to be contained in one shot, it sometimes helps to look for symmetry. As in most other photography the best time to catch the beach patterns is early morning and late afternoon, when the light is kind and not too glaring and when shadows define ridges and the pellets the crabs make.

I have used many kinds of film but Tri X more than anything else, and usually at 800 ASA. Color—Kodachrome X, Ektachrome X, Agfa CT18, and sometimes Fujichrome. Color I do not process myself, but with a black and white I often try to get a grainy negative by overexposing, processing with a print developer, and so on, as grain accentuates the sandiness. Sometimes exposure is tricky, for you are on a beach where the light tends to be strong, and at the same time you are usually directing the camera downward so that no sky or sea shows. Unless you are an expert judge of light, use a meter and overexpose a stop or two with black and white. With color, overexpose less or not at all for closeups; for long shots, I generally underexpose.

Step lightly and look where you're going or you may spoil a pattern and miss a photo.

Introduction

Vision opens up the world for us. With it we can reach beyond the boundaries of the tangible world, bringing into simultaneous focus broader and more complex events than can be perceived by any other sensory means.

Today, new instruments and associated techniques have extended the spectrum of the visible world. Partly by substituting the camera for the human eye, partly by using new means of transportation, from airplanes to rockets, observations can now be made at hitherto impossible angles, from inaccessible places, and under physical conditions that would destroy human beings. The permanent record that the camera provides gives us an opportunity for sustaining visual experience as long as we wish, long enough to overcome errors that the eye makes because of our impatience, prejudice, and inability to recall. These newly extended vistas have expanded man's knowledge of his world, his control of its energies, and his use of its configurations.

But more than giving us merely useful information about the forms and processes of nature, this new vision has given us the possibility of revealing the patterns of nature. Every pattern in nature is like a secret sign, a hieroglyph of nature's hidden messages, which may be seen only by those for whom the world is full of mystery and who search for the key to unravel it. Some poets and artists have this key, as do some scientists who are also poets. Emerson and Thoreau had it, as well as Paul Klee and D'Arcy Thompson. What they have in common is the ability to look at things in nature with the wide-eyed wonder of a child and see patterns most of us would miss, reminding us (and we have to be reminded) that the variety of patterns in nature all belong to a single, all-encompassing pattern of "nature's community."

Far Eastern philosophy grew out of the notion that man lives most fully by opening himself to the universal rhythm of nature, by becoming one with the trees, stones, and animals. What is important is not seeing the world in terms of likeness, but in terms of what the Chinese call "rhythmical vitality," or the "essence of things," in their characteristic life movement. Stella Snead's artistry is akin to this Eastern vision; her photographs pulsate with the pure rhythm of undisturbed nature. With absolute control of her craft, and poetic perception like that of the great visionaries mentioned above, she succeeds in transforming nature's smallest details into life-affirming micro-miracles.

This is needed assurance in our often confusing, man-shaped world.

Gyorgy Kepes
Director, Massachusetts
Institute of Technology Center
for Advanced Visual Studies

When we go down to the low-tide line, we enter a world that is as old as the earth itself—the primeval meeting place of the elements of earth and water, a place of compromise and conflict and eternal change. For us as living creatures it has special meaning as an area in or near which some entity that could be distinguished as Life first drifted in shallow waters—reproducing, evolving, yielding that endlessly varied stream of living things that has surged through time and space to occupy the earth.

—Rachel Carson

Take a beach, a wide one, lonely, the tide far out. Not a specific beach, yet one in a tropical part of the world, a beach that is really the memory of several beaches fused—the memory of beaches known to me because I lived beside one of them for eleven years and paid frequent visits to the others, in fact, to many others around the world, northern shores included. Provided these beaches are deserted and sandy there are more similarities than differences in the patterns that form on their surfaces.

Let us say the beaches in this book run down the west coast of India from near Bombay on southward. The Arabian Sea washes them and the water is usually warm. On the land side, there may be palm trees, feathery-leafed casuarinas with sturdy reddish brown trunks, or a barrier of cactuslike hedge.

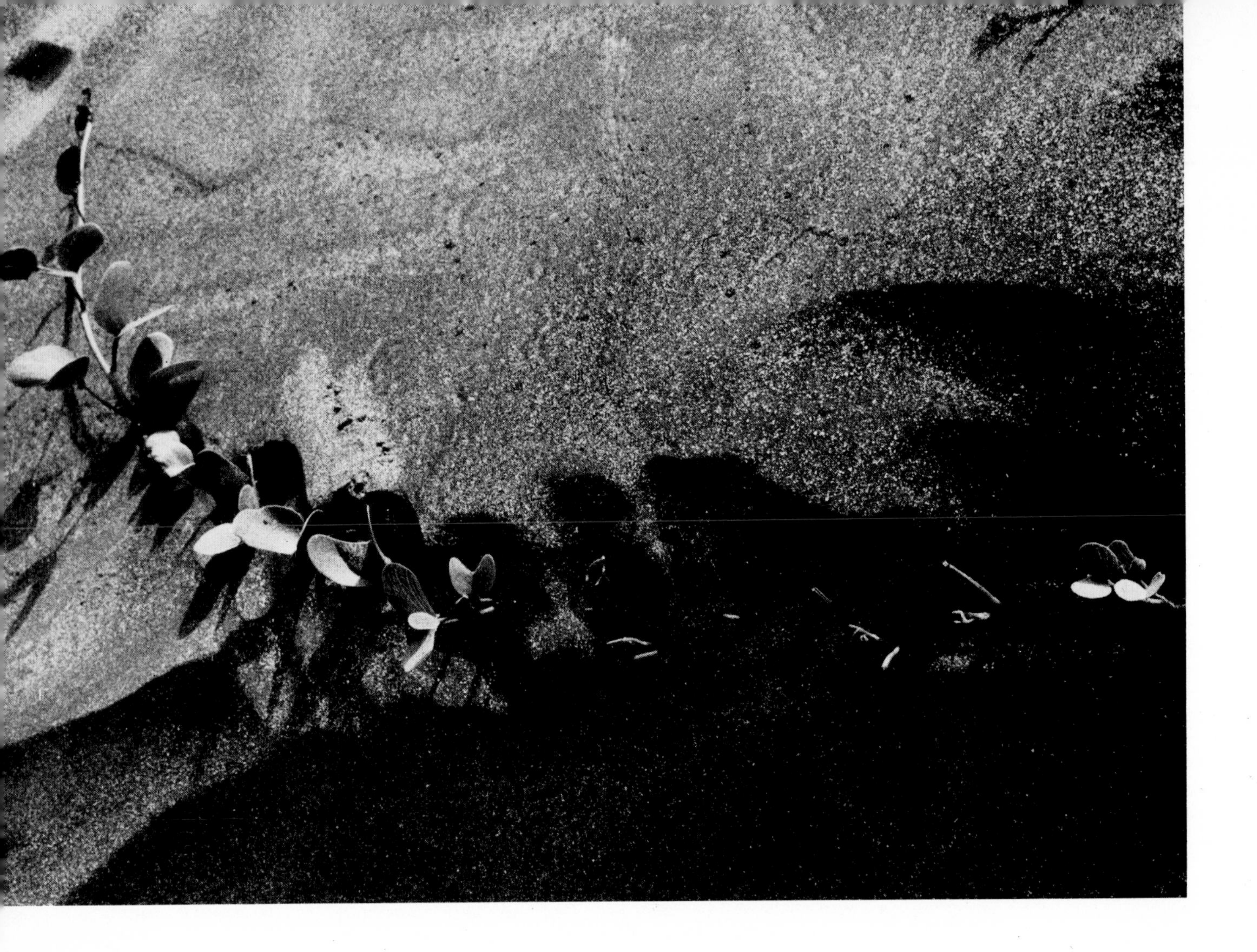

On the soft, dry upper sand, a trailing plant with thick wing-shaped leaves appears to live a rambling and contented life off the sand alone, sometimes gathering itself into clumps as it spreads across the duny landscape. In places, a discreet house may be seen through the higher foliage, or a fisherman's shack, or, if a city is not so far away, a tall building bursting above the trees, aggressive and unwanted. Even when a city is nearby, such a beach can still be deserted most of the time. Visitors from abroad are few, and the local people come in crowds only at sundown on Sundays. For a brief hour or so they cover the strand with colors as they stroll in the cool breeze of sunset. At dusk they scatter and leave the beach alone for another week.

A distance offshore is a wooded island or a rocky one with a fort; one such island can be reached by walking across the mud flats at low tide. The beach stretches invitingly. Only one figure can be seen, but perhaps there are others bending over rocks, searching in pools. Suddenly the quiet mood is broken as two carts drawn by bullocks sweep by, racing one another, their drivers standing, wielding whips, and yelling. They are quickly gone and tranquillity returns. At one end of the beach a low headland curves and from behind it small boats emerge with sails set. They sail to the horizon and then along it in a neat line. The sea is calm, not very blue; small waves pleasingly turn and flow unceasingly inward and outward.

On some of these beaches the sand is striped—dark, light, dark, light—giving the illusion of ridges, but we who have walked there know the surface to be perfectly smooth. The stripes fly off before us in sweeping curves and zig-zags. They remind us of watered silk. Made by the water with its to-ing and fro-ing, uncovering and covering the dark sand with the light, they are the ripple marks of the waves whereby the ocean almost succeeds in making the beach resemble itself.

In most seasons it is a mild sea. It does not throw up rough waves to crash among pebbles, to shatter shells, or to retreat with a clatter as on steep, stony

shores. Rare is the wave that splashes high against rocks, displaying itself in fury or joy. And there are no cliffs.

What is there then about these halcyon expanses of sand that brings us back again and again, at all times of day? The answer is to stop looking only ahead or to left and right, eyes level with the horizon, but to look down. We may start this immediately on emerging through the gap in the hedge early in the morning. There, stamped faintly in the dew-damp sand, are the footsteps of a mongoose. This mongoose did not venture far from the shelter of the vegetation. It is as if he ran and looked and knew that these great spaces were not for him. After nights of heavy dew or when it has rained, drops of water shine on leaf and twining stem before being scorched away by the day's heat. The sun rises behind the trees so that at first the upper beach is somber, long shadows tapering down toward the sea. The sand is cool to the feet, not burning as it will be later. Soon the slanting rays break through, and a section of small dune is lit. Photographed it looks like the Sahara. If we bend down, lie down even, we can see the same curves of sand against a vast sky, a shadowed valley, bumps and hillocks. And why not? Sand has a certain uniform behavior. If it is dry, its contours are formed mainly by the wind, the particles adhering or falling in the same manner whether the expanse is great or small.

The composition of sand began eons ago, after the first rocks had formed on the earth. With their gradual weathering and decay, chunks and particles fell into the rivers, there to be further broken up by the action of water. There were rockslides too, and sections of riverbanks fell in, all joining the surge of sediment to the sea. In time the sea washed some of it back onto the edges of the land; we call these edges a beach.

A sandy beach is much older than a rocky one, and sand, surprisingly, is far more durable than rock. Each grain is almost indestructible, a minute fragment of the hardest minerals surrounded by the unlikely armor of an infinitesimal layer of water, adhering by capillary action and guarding the speck from further weathering. Beach sand consists mainly of quartz, but there are many other minerals present, even semiprecious ones such as garnet, tourmaline, and magnetite. Lightweight minerals such as mica do not sparkle for long on a beach as they are quickly carried away by wind or water; the heavier, coarser ones, like ilmenite and rutile, remain and make up the black sands that streak some beaches. The strands bordering volcanic islands are often all black, the sand being derived from basaltic lavas that once welled up from the interior of the earth. There are beaches made up of shells and coral debris, shattered, ground, and polished by the waves to glisten and fascinate, but here and now we are concerned mostly with light quartz sand and the coarser, darker variety.

Rachel Carson expresses the primeval nature of sand in these words: "And the materials of the beach are themselves steeped in antiquity. Sand is a substance that is beautiful, mysterious, and infinitely variable; each grain on a beach is the result of processes that go back into the shadowy beginnings of life, or the earth itself."

Sand—or the mountains of Persia. Seen from the air the contours show plainly as on a map.

At the high-tide line there is the usual spread of cast-up flotsam: shells no longer inhabited, dead fish, broken coconuts, wisps of seaweed, parts of a lobster shell, and bits of wreckage.

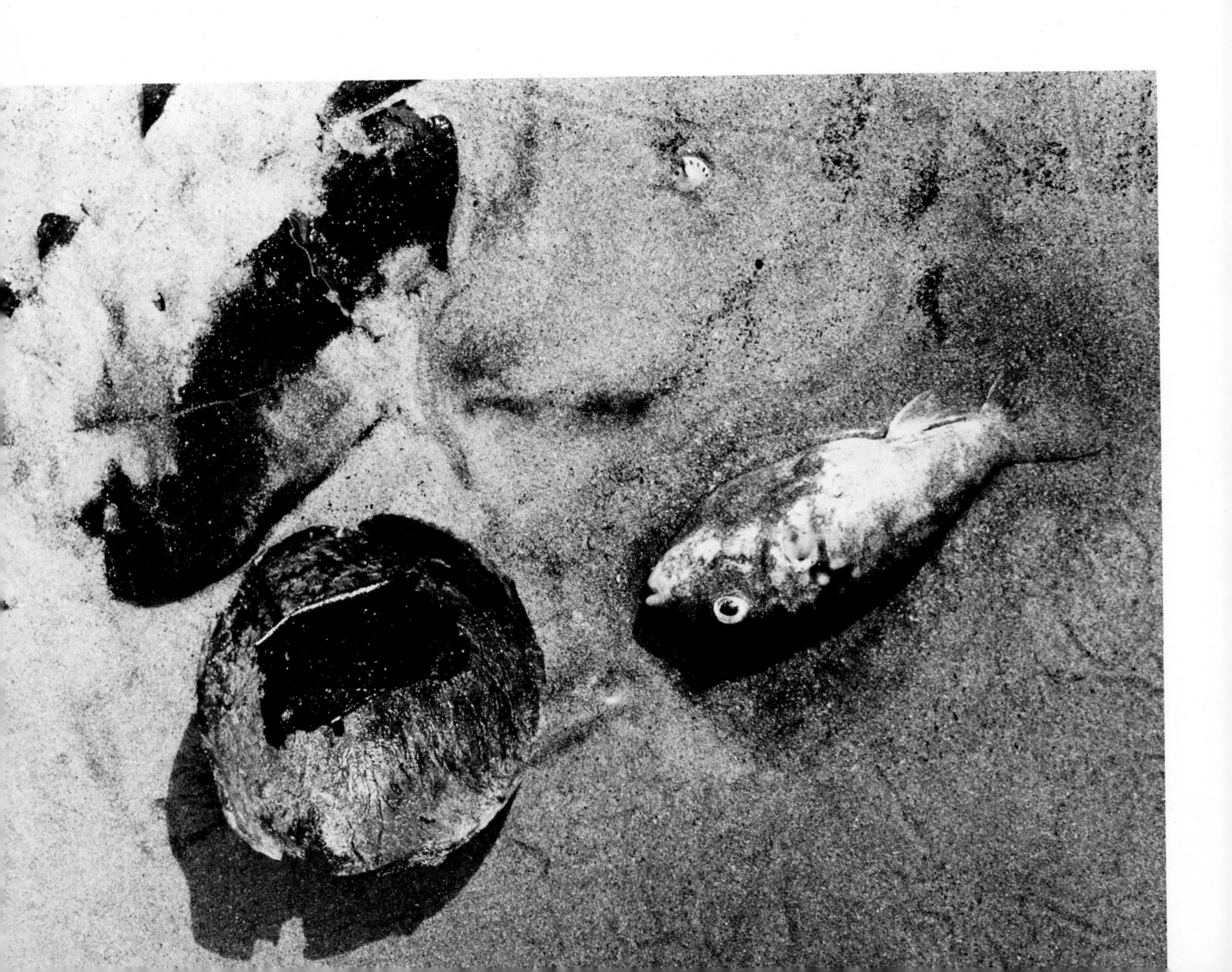

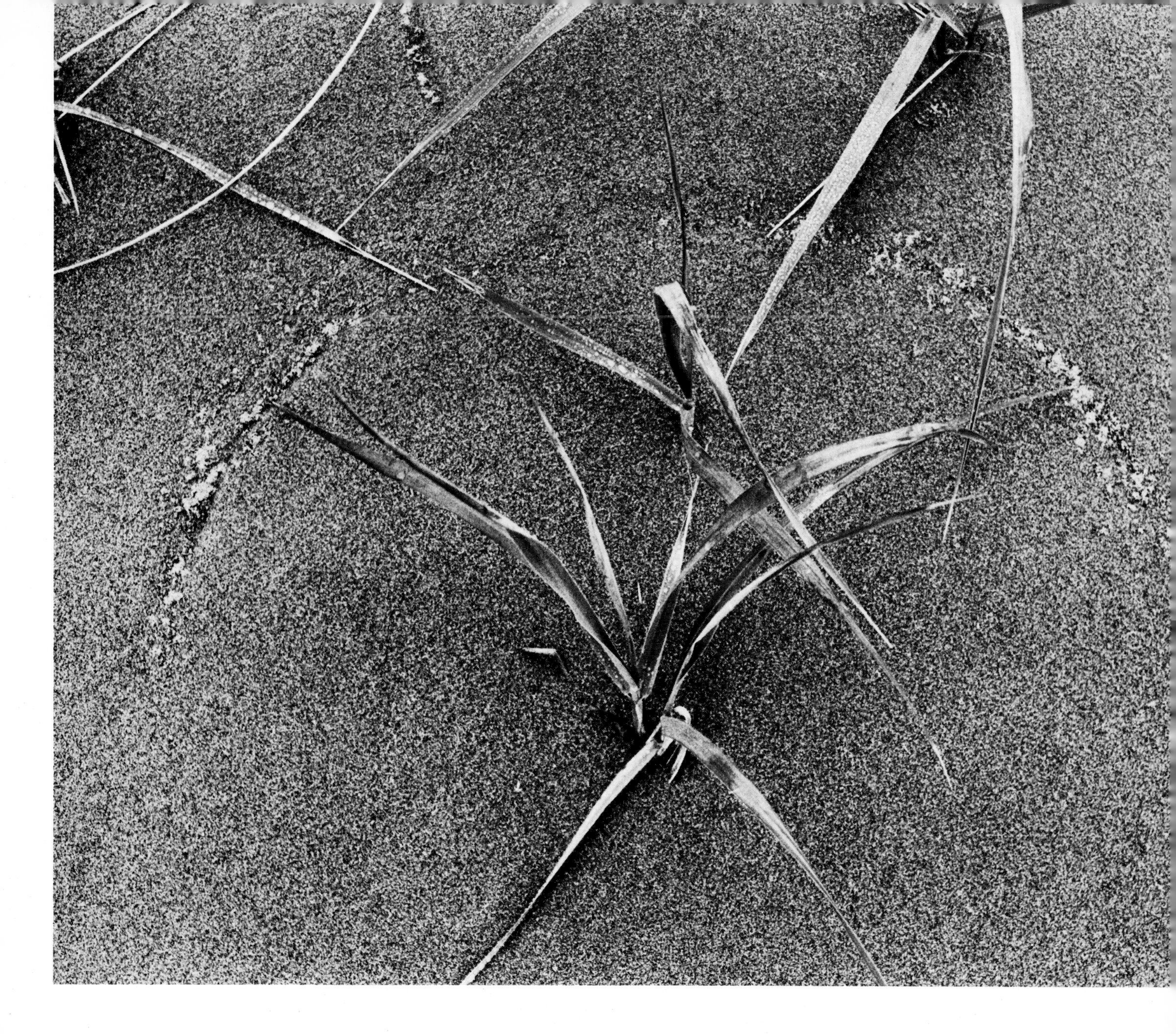

Neat plants sprout a little removed from the debris, their knifelike leaves swing in the breeze, dipping their tips to carve circles and semicircles in the sand, their points the pencil, their center where they root the compass point.

Lower down on the flat, newly washed sand, intense activity has started. Thousands of small white crabs are scurrying out of holes and back again. At each brief appearance they deposit a tiny pellet of sand. As we approach they disappear, but once we pass work is energetically resumed. Their method of feeding is such that they must suck and swallow a great deal of sand in order to extract the particles of nourishment that exist both in the sand and in the minute casing of water that surrounds each grain the sea has touched. The little balls the crabs throw out with such meticulous dispatch are waste; here are creatures who really mean to keep their homes clean and uncluttered. Every moment the pellets accumulate and spread around each hole in an endless variety of patterns. Sometimes the pellet patterns are so many that they become joined to form one immense carpet that we must walk on; the design then is found in the occasional empty space still free of pellets.

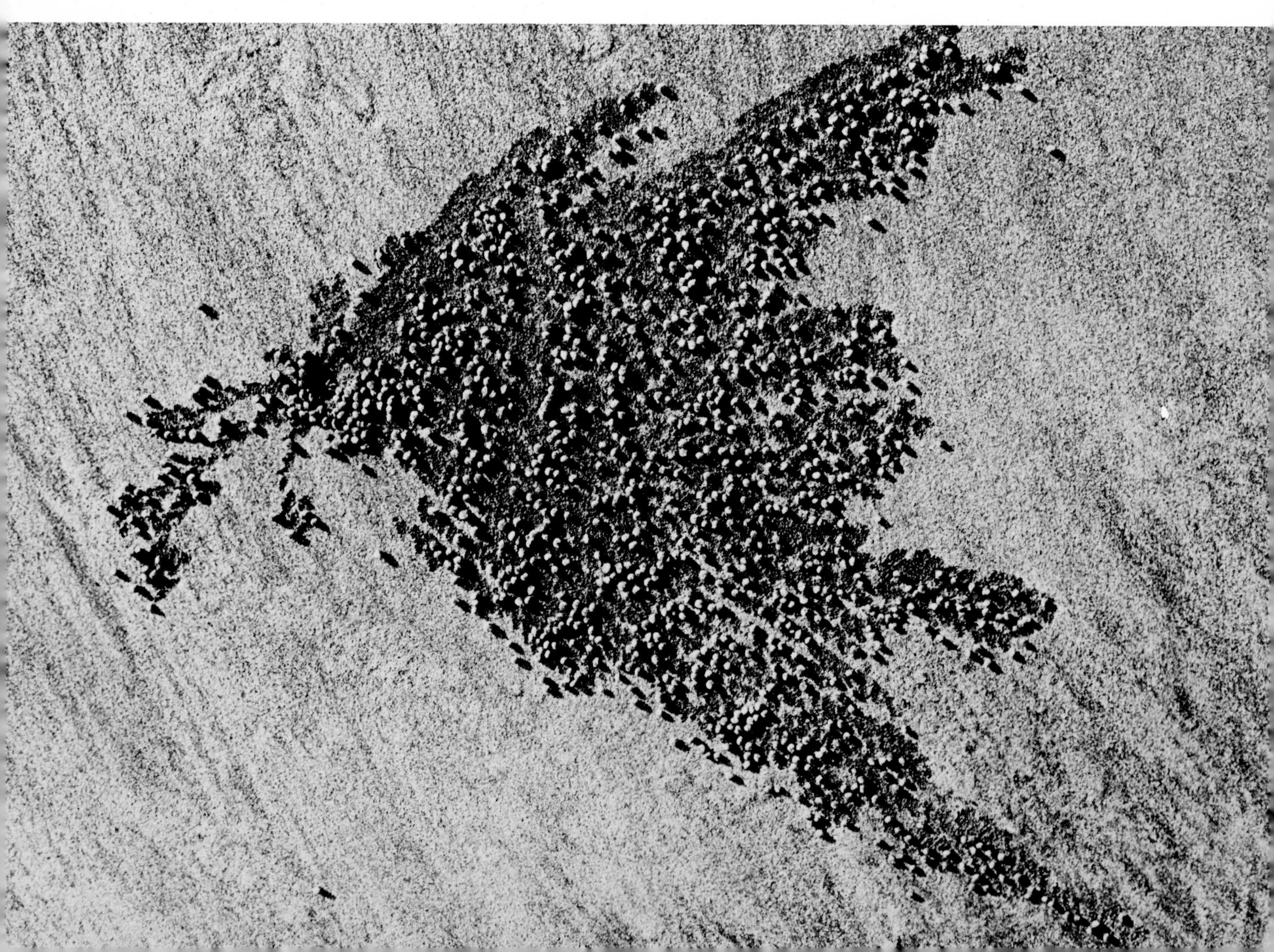

But usually the crab population is less dense and the holes are more thinly scattered so that there is an area of plain sand between each pattern. Never was there a more unconscious art, nor one so satisfactorily devoted to chance. The results are never the same, and they are as unexpected as they are fascinating. Here is a flower with a stalk and a leaf attached; here a butterfly with folded wings—the crab's butterfly, for you can see three disappearing crab legs. Shapes with rays spreading from a center, the hole, are common, but how did this one become so regular, like a quick drawing of the sun or a star?

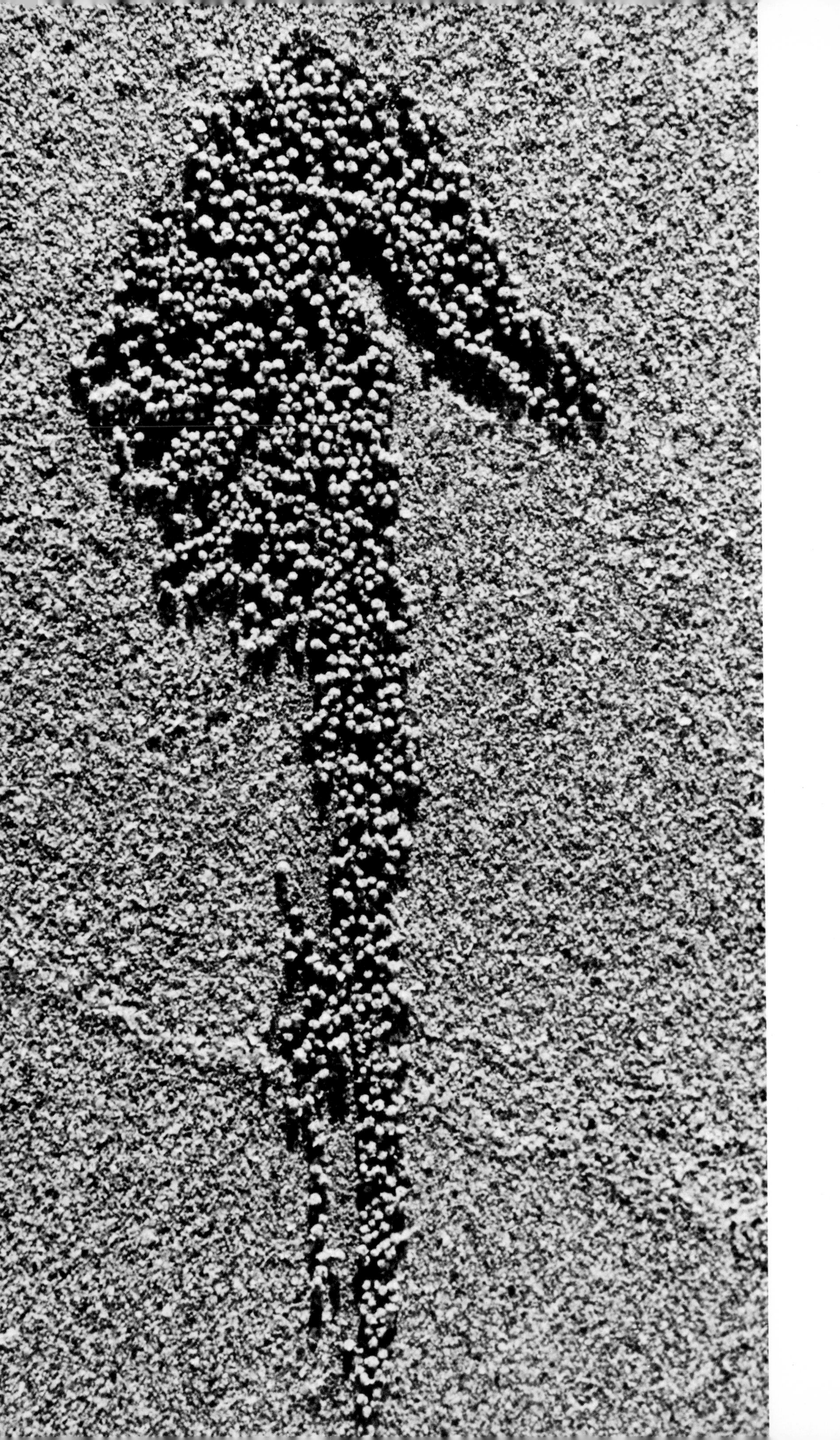

This grouping in the form of a seahorse was about three inches in length. When I enlarged the photograph to 20″ × 24″, I saw something else, something that might have been seen from the air: a herd of animals gathering round a waterhole, an oasis of trees, a crowd of people under white umbrellas. One does not often think in these terms while on the beach; it is upon examining the enlarged photographs later that these fancies suggest themselves. Then there are the seemingly cooperative efforts of a number of crabs.

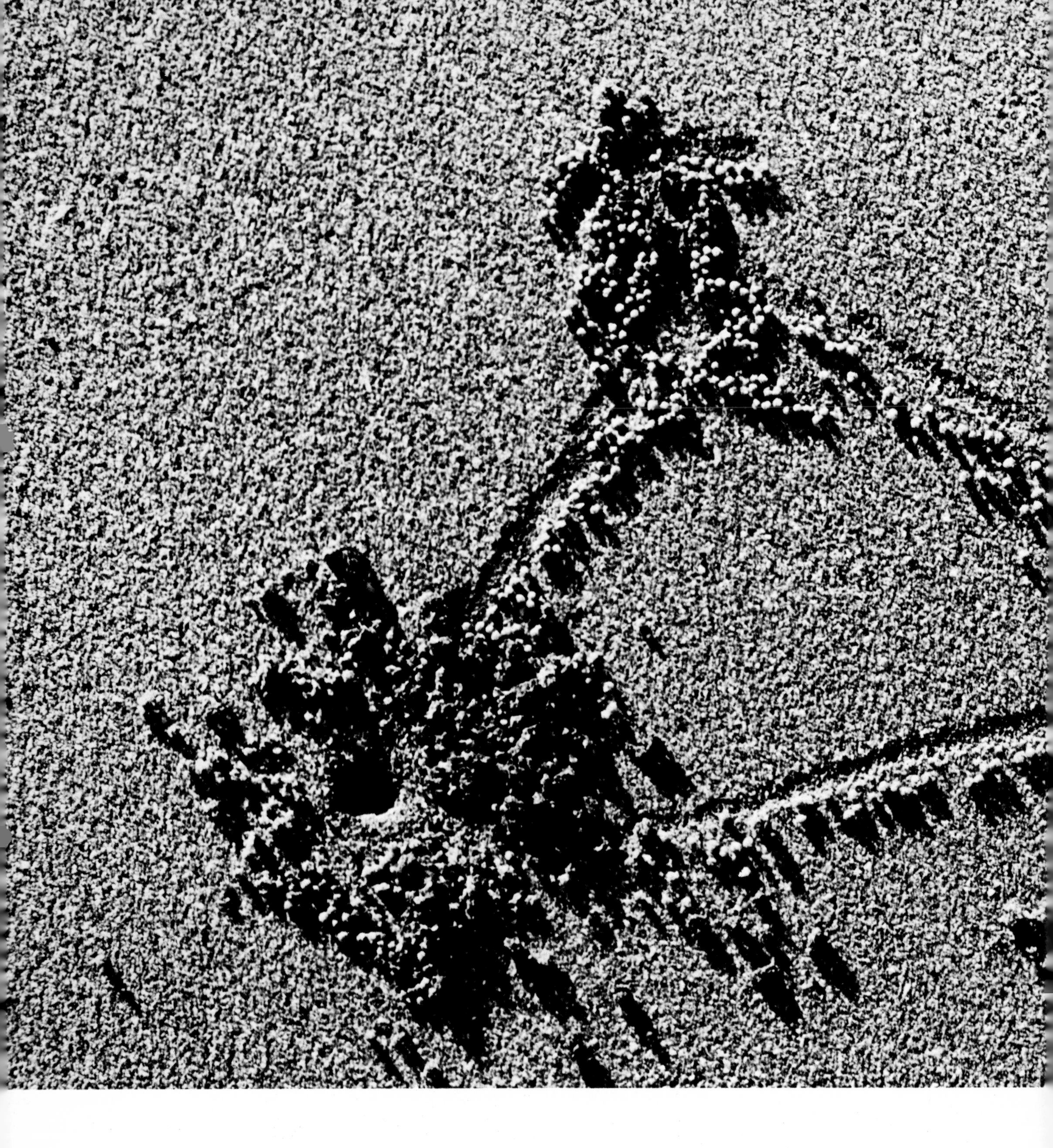

Although the idea of a collective invention is our illusion, there are, nevertheless, sometimes distinct and direct tracks between one hole and another.

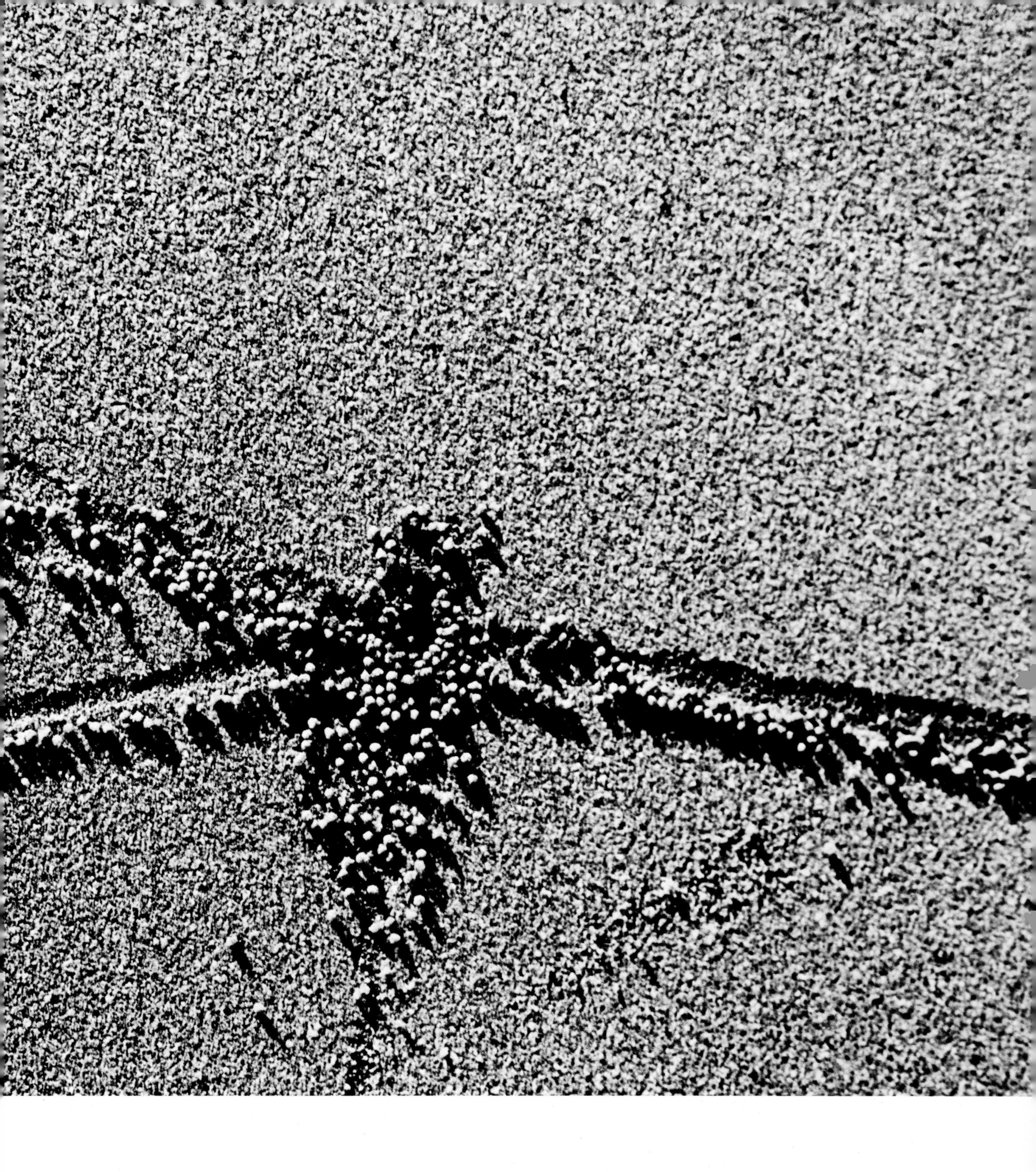

Enlarge the photograph and again we are in the air, this time looking down on villages with tree-bordered highways running between them.

These small crabs are the neatest. Their pellets are always perfectly circular balls that never fall apart no matter how hurriedly they are dropped. Larger crabs seem to be more careless and untidy. They make larger holes, but the debris around them does not make pleasing patterns; in fact, it resembles the mess a human makes when he digs a hole. Twice in every twenty-four hours all the visual results of this singular activity are completely blotted out by the rising tide. The crabs retreat beneath the surface, while the sea replenishes the sand with what is needed for their survival. And so, very soon after the waves recede leaving a pristine flatness, the first crabs bob up and the whole process repeats itself as it has from the very beginning, in darkness as well as in daylight, and with never a single pattern exactly like any other.

Farther down the beach, beyond where the crabs work and where by now the sand is relatively dry on the surface, is a damp and shining area. If there are rocks, they usually begin here too; they are low, fairly smooth, easy to walk on. Or there are scattered boulders girded by pools. Occasionally one such boulder is crowned by an object like a golden diadem; it is a welk egg-mass wherein the eggs, each one in a little bag-shaped capsule, are protected from external injury.

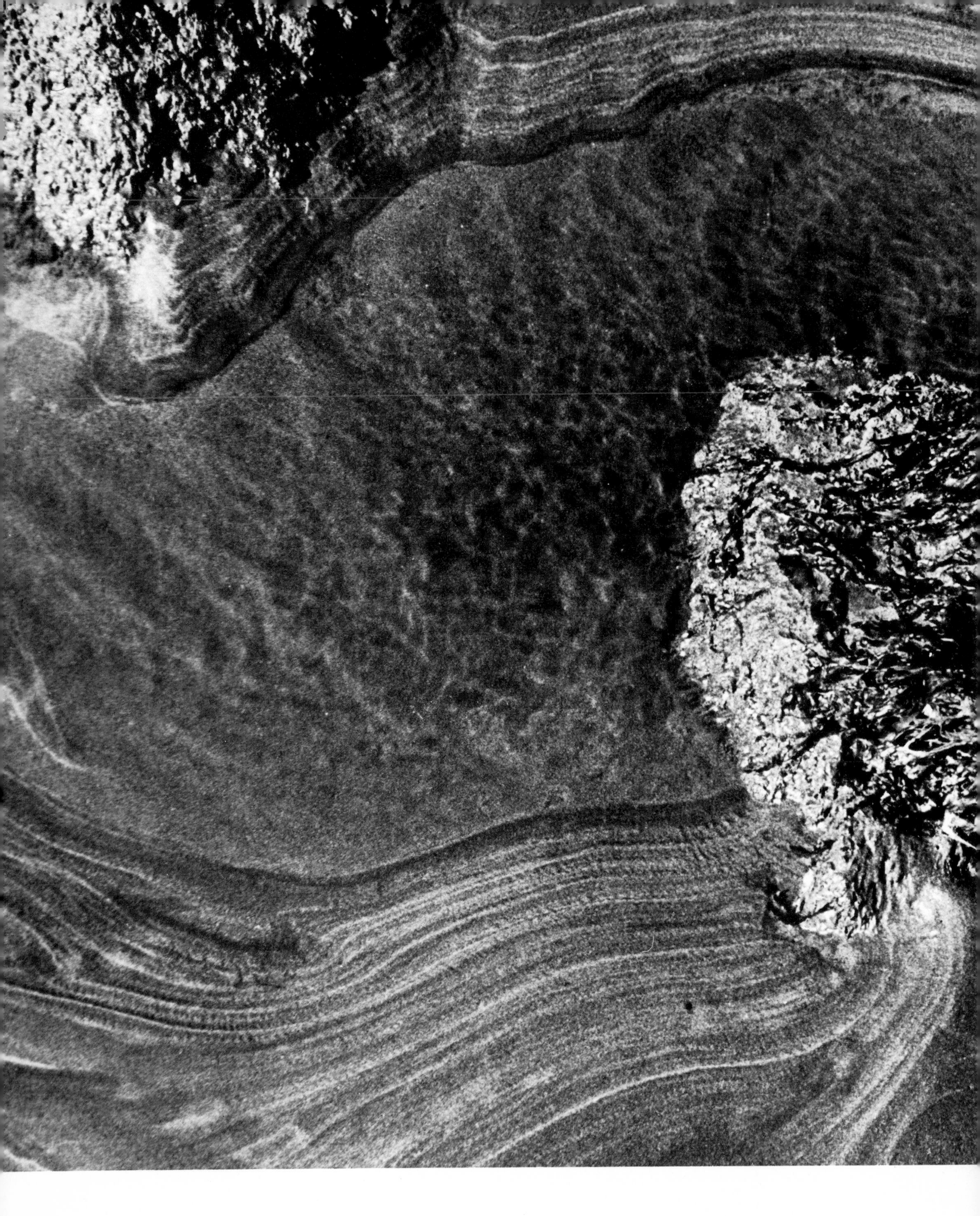

Some of these rocks and gently banked edgings of sand may encircle a pool.
Seen from above, from a person's height, it could be a coastal landscape,

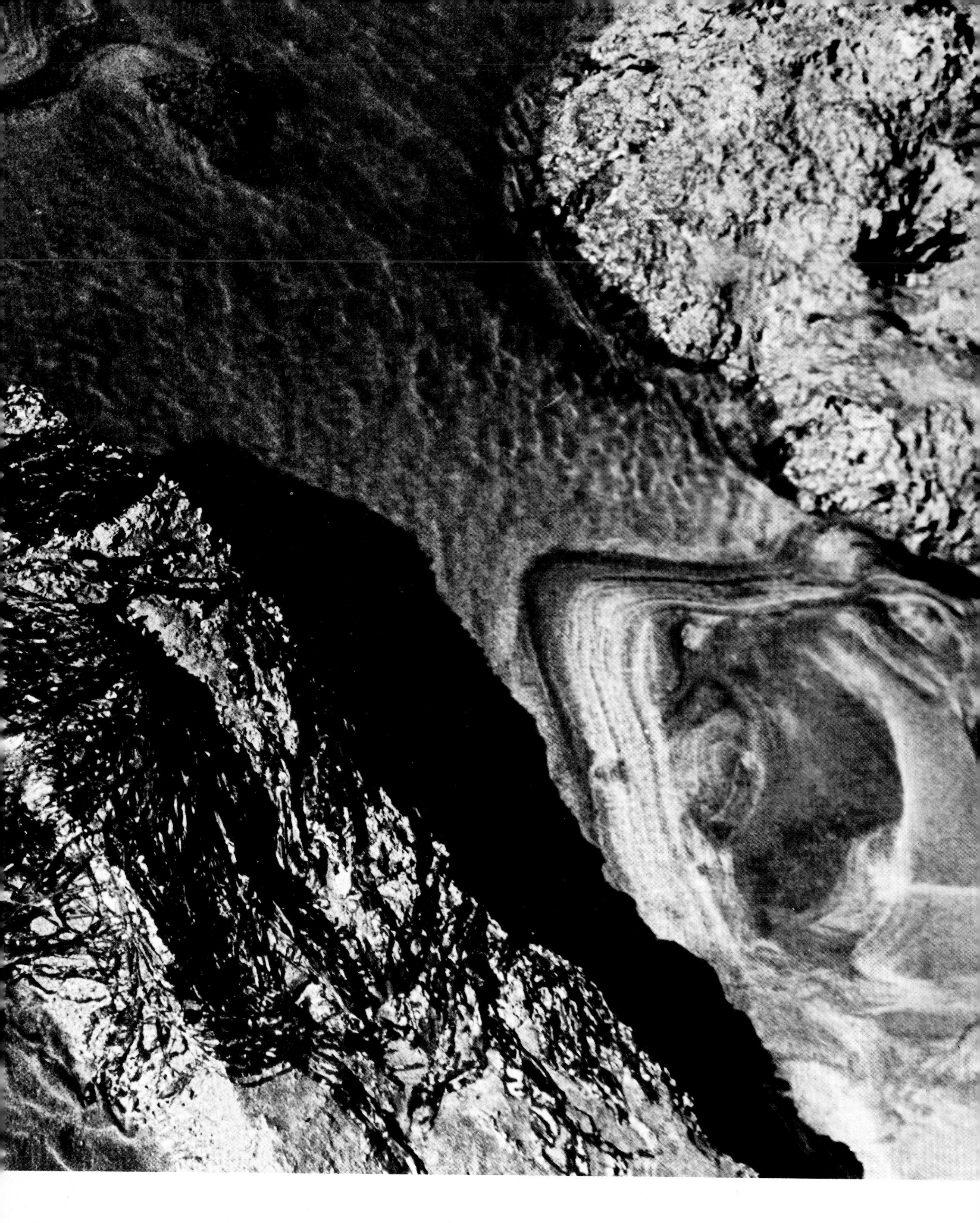

rocks jutting as headlands, the sand elegantly lined and curving as luxury beaches.

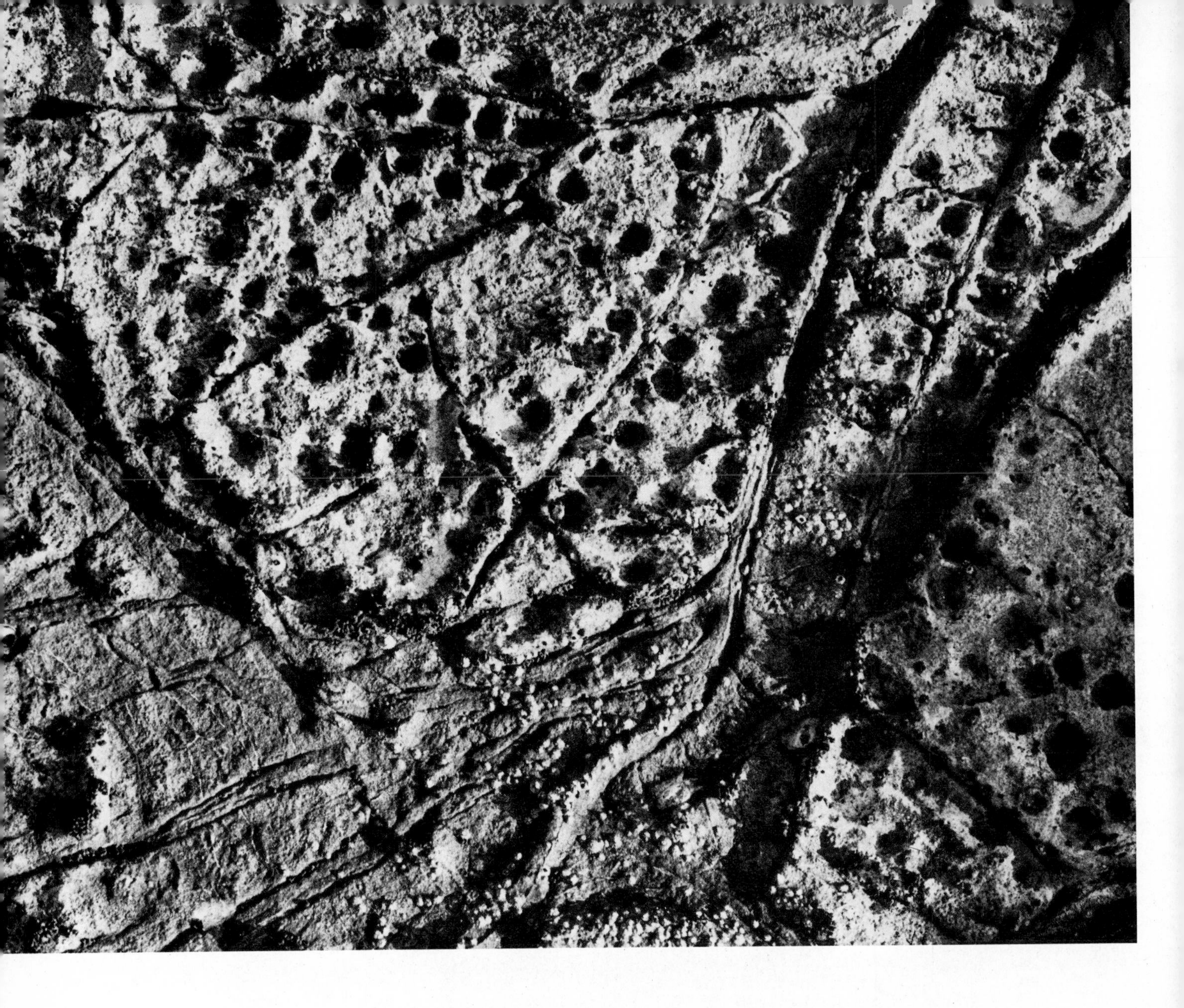

Certain rocks are pockmarked by rock-boring bivalves, which, it is thought, are aided in their efforts by being able to emit an acid; other rocks, carrying seaweed and barnacles, stretch for some distance; all are covered daily by the tide.

In between the rocks and around their bases, the water swirls, leaving as it ebbs patches of sand, now smooth, now rippled. In keeping with the relative quietness of these seas the swing of the water is delicate, but it is strong enough to etch enchanting embroideries in the sand. They resemble pen drawings rather than action paintings, yet there is nothing cramped or stilted about them. They are free flowing, for what makes them is a flow, a trickle, a creeping of water draining from the sand. Around the base of a rock there is a pool and into it the trickles wander, slightly indenting the sand, making runnels that branch from and into one another just as great deltas do as they approach the sea. Again, around a rock the retreating water has circled leaving long tresses of pattern; there are confluences and dry endings. Should there be darker sand just below the top layer, the patterns are strikingly accentuated. To photograph them, one should be there soon after the tide recedes, when edges are clean cut and no foot or dog paw has marred them.

Similar tracery occurs on open stretches of wet sand. Here the shapes are more treelike and they are arranged in regular rows, the runnels always at right angles to the sea; at one end they branch like trees, at the other they

billow like cloth. When in rows these "forests" are about a foot in length, but frequently a stronger runnel will furrow its way on and on down the beach for several yards, weaving and wandering and folding over itself like strands of long hair.

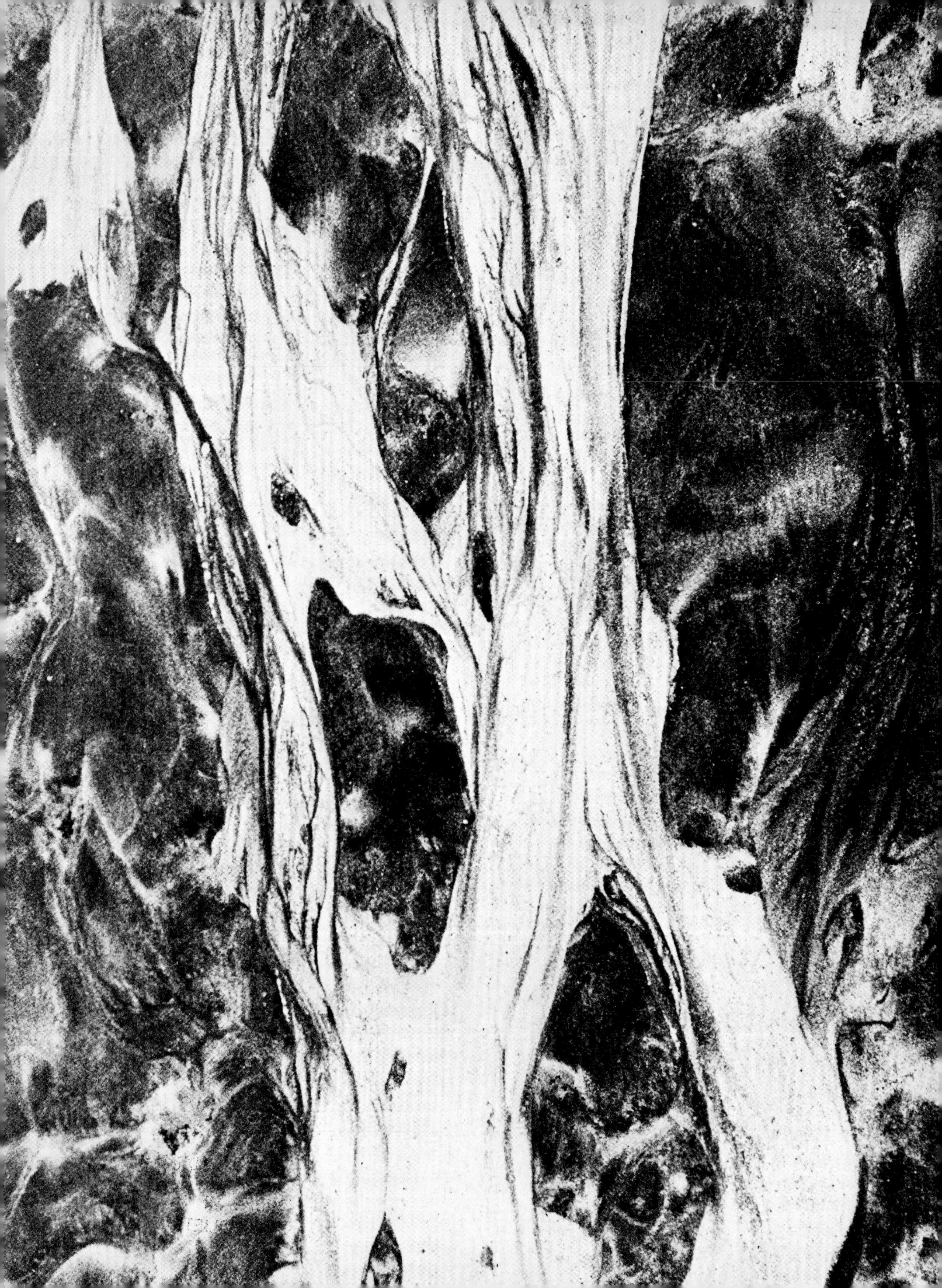

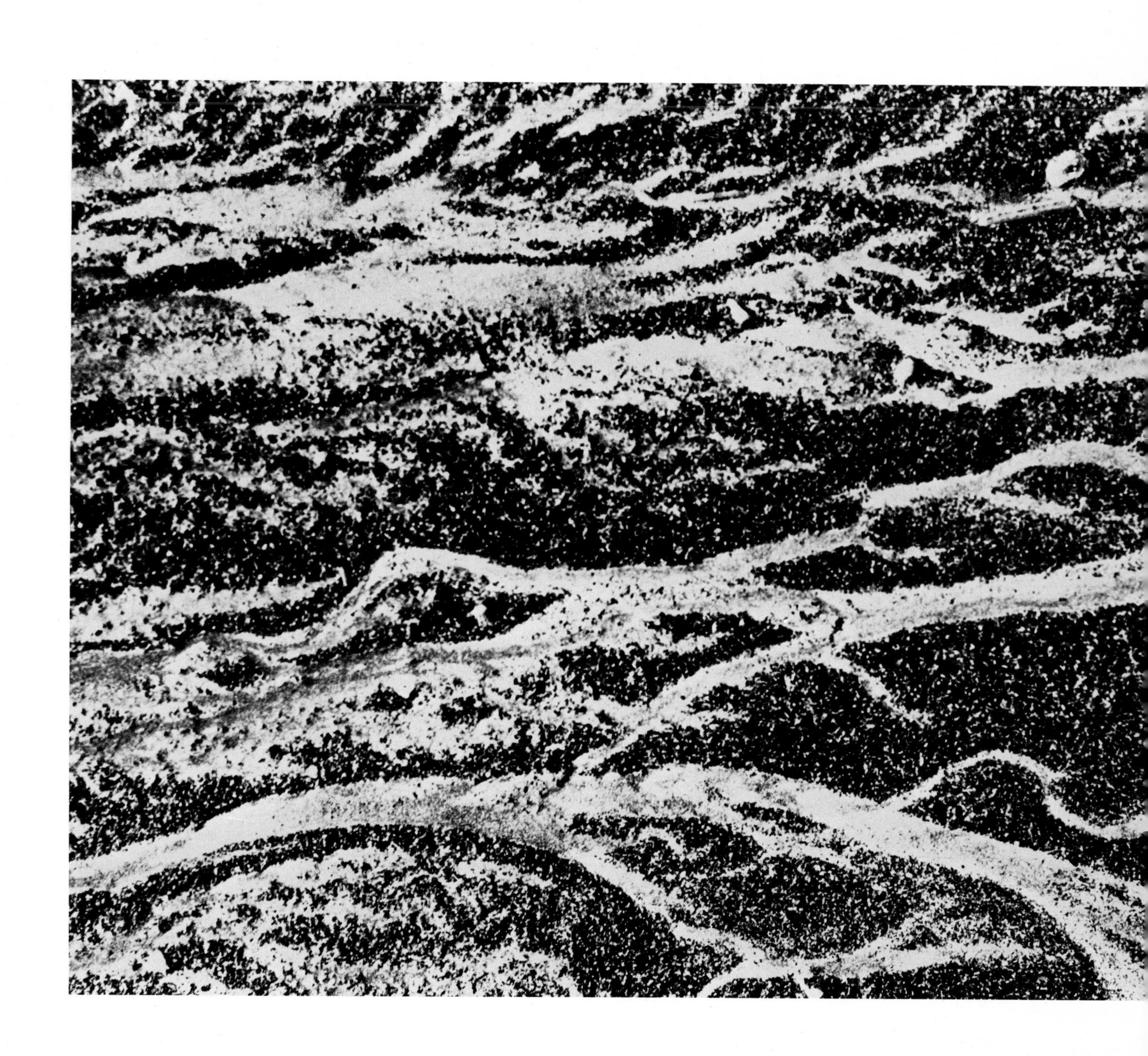

Another caprice of the water is to drape a layer of light sand, looking like a piece of torn cloth, over a dark area. The light sand weaves freely among the dark, yet because of their different textures the two kinds of sand lead more or less separate lives; thus the patterns last for quite some time.

Besides these water and sand configurations there is much else to observe on the lower beach. It is here we realize there is more going on beneath the sand than above it, and now that the waves have retreated the clues to the underworld begin to show. There are holes that bubble, tiny protruding chimneys, and little pulsating bumps, some of which move along in a confusion of rounded lines. And there are tracks. Some, of course, are of birds; the webbed feet of a gull or three prongs and a dot signifying a crow, a frequent visitor that pecks at broken coconuts and other debris. Also on the surface are marks like tank treads running resolutely across the sand and around the base of rocks. If we are quick enough, we might see the crab that made them disappear into a seaweedy pool.

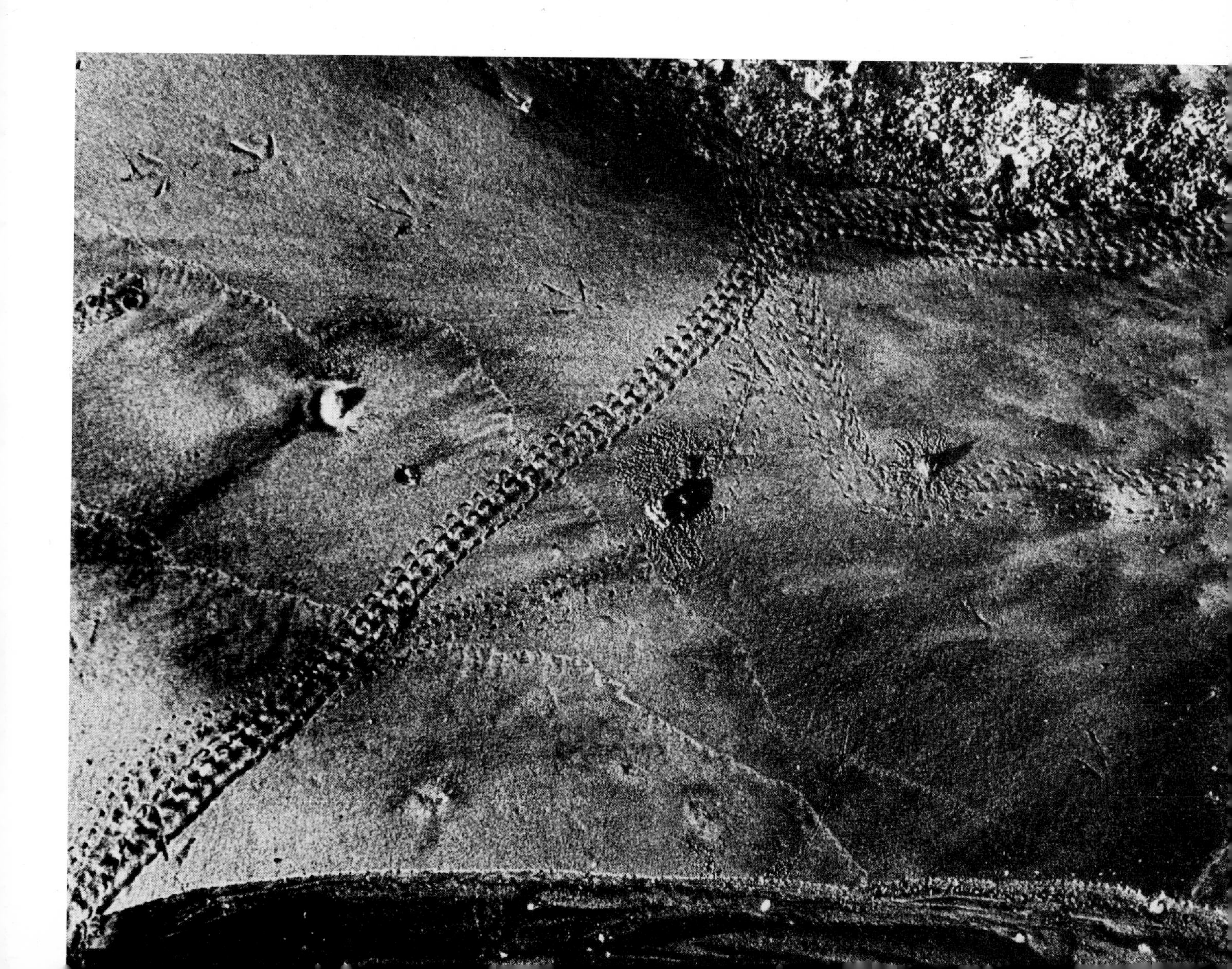

Certain formations are the result of the doings of an unseen marine creature, one, for instance, which buries itself under a small mound, which in turn causes plantlike streamers to appear in the sand as the water seeps by toward the ocean.

But much more common are the trail patterns made by food foragers moving just below the surface of the wet sand. Like a child's scribble, they cross and zigzag in seeming confusion.

Or they branch in a more orderly way like a plant. Watch the lacy embroidery develop as its unconscious designer, perhaps a mollusk under a small, rounded shell, searches this way and that for food. It would seem that certain shells cannot resist attaching themselves; on the opposite page they have selected, for want of anything better, the rubber sole of a discarded sandal.

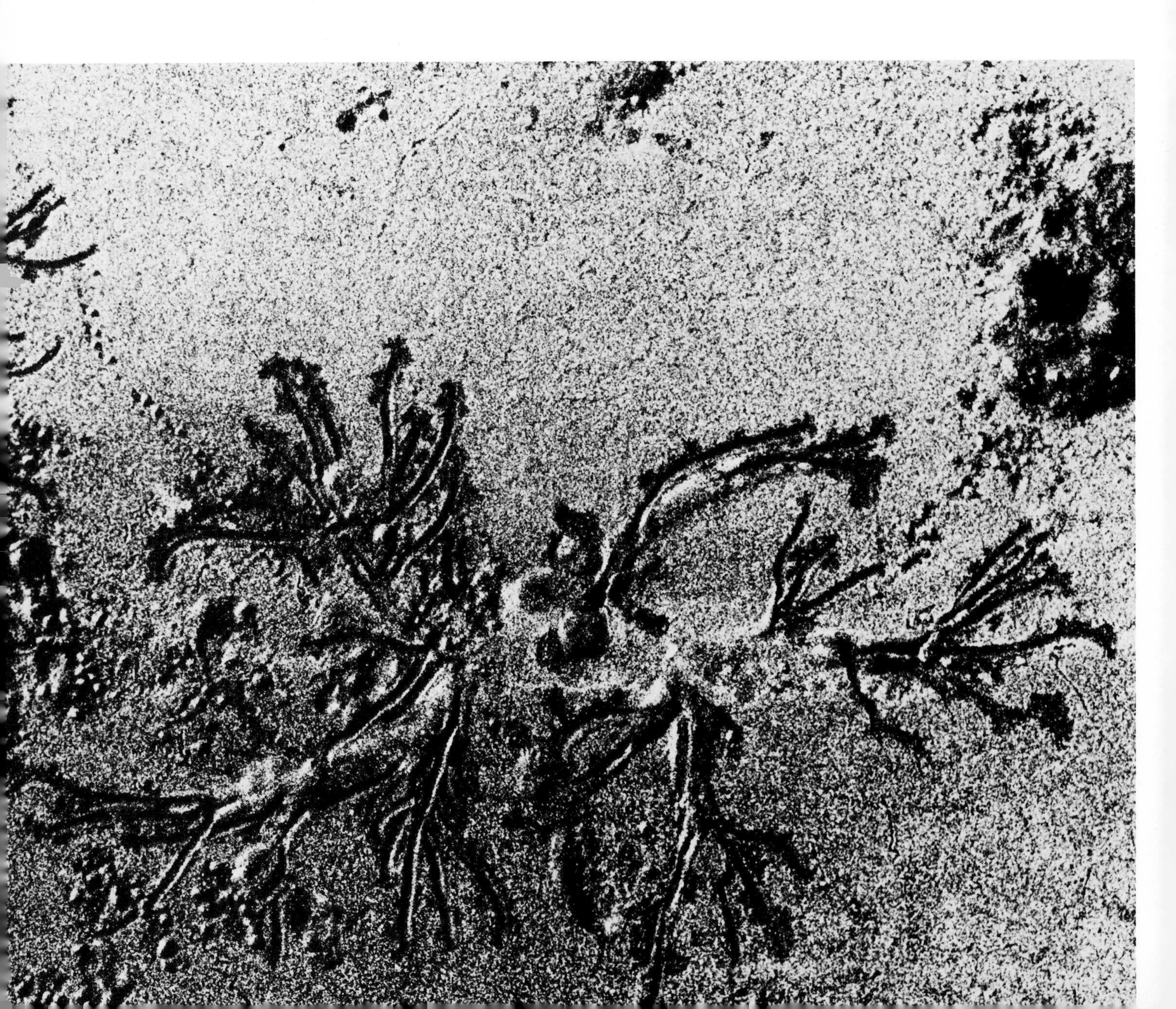

Then again, as on the following pages, there may be a frenzied-looking scamper for food resembling an aerial view of a tortuous mountain road. There are also shell-covered searchers that deem it unnecessary to have a protective covering of sand at all times. They push around in a tangle of trenches. If danger threatens, they duck under the sand and the canyons end abruptly in circular mounds.

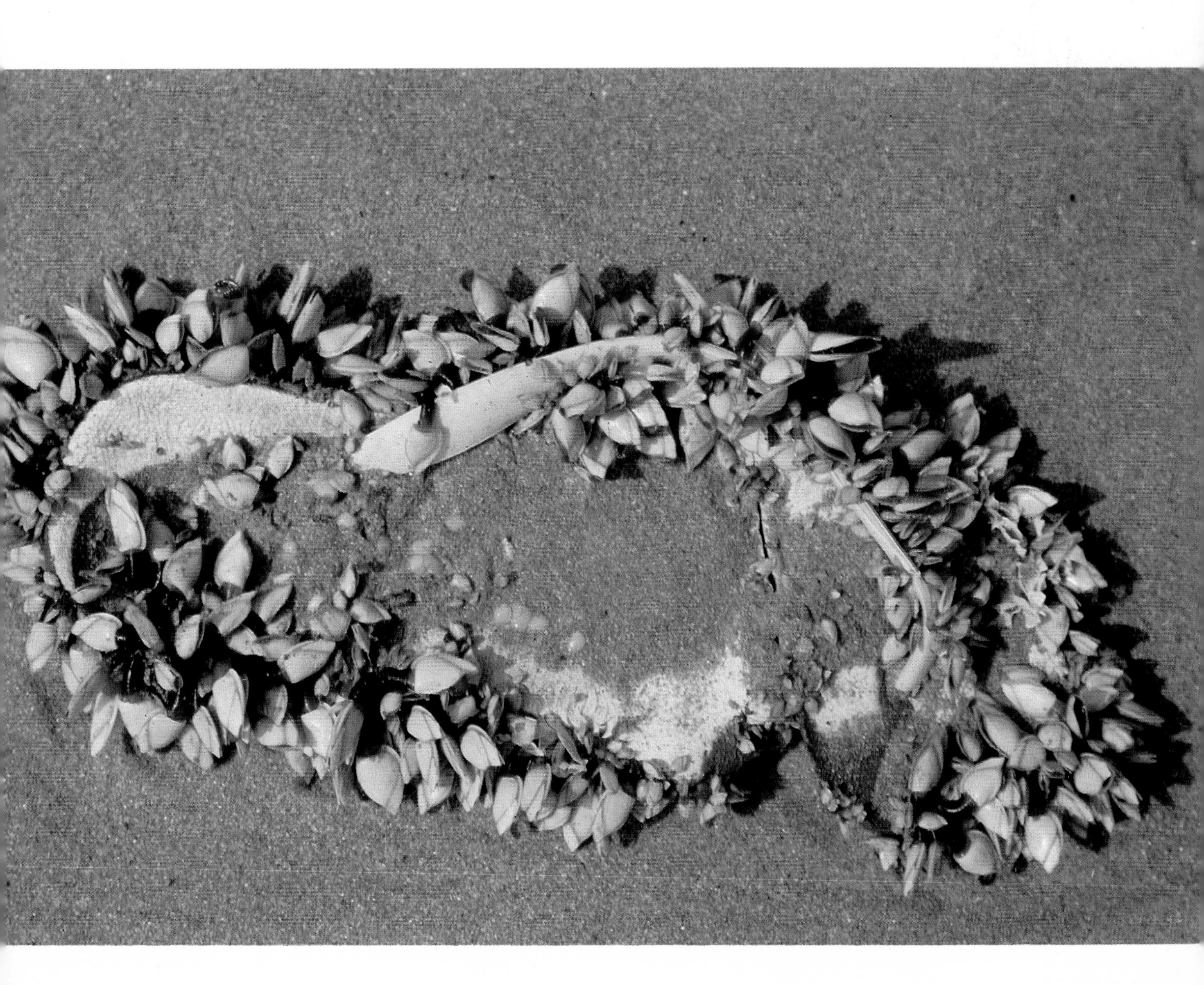

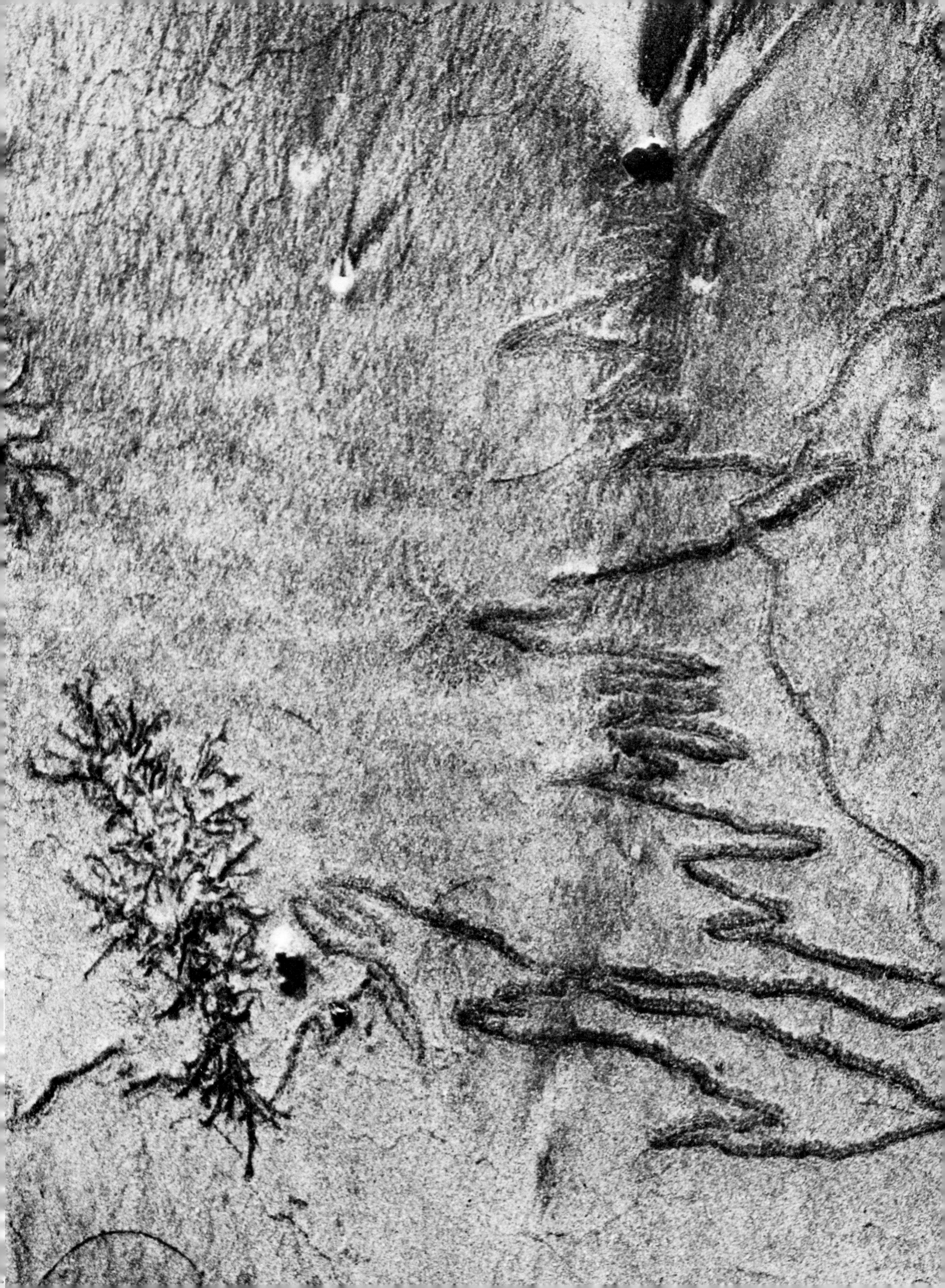

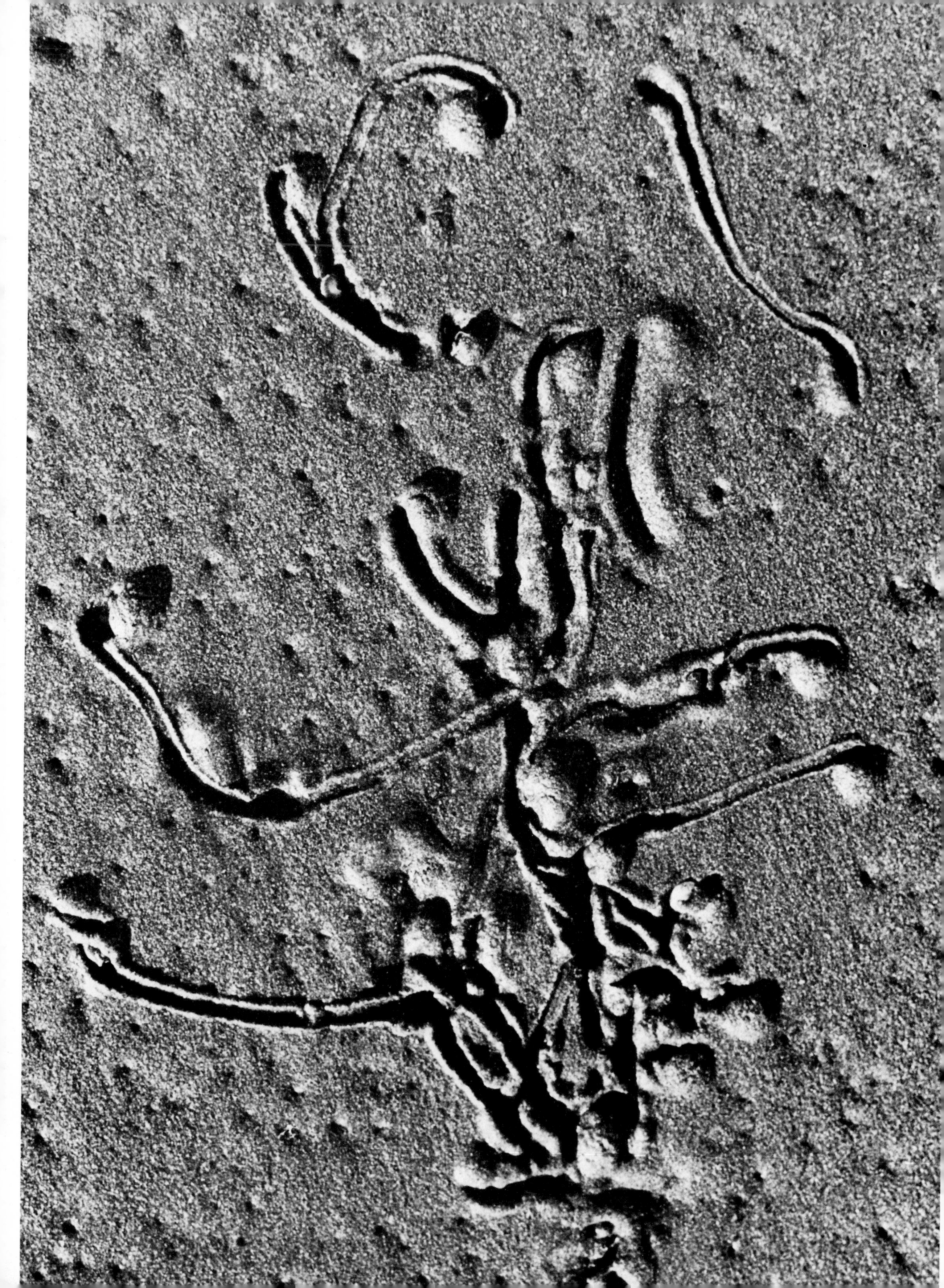

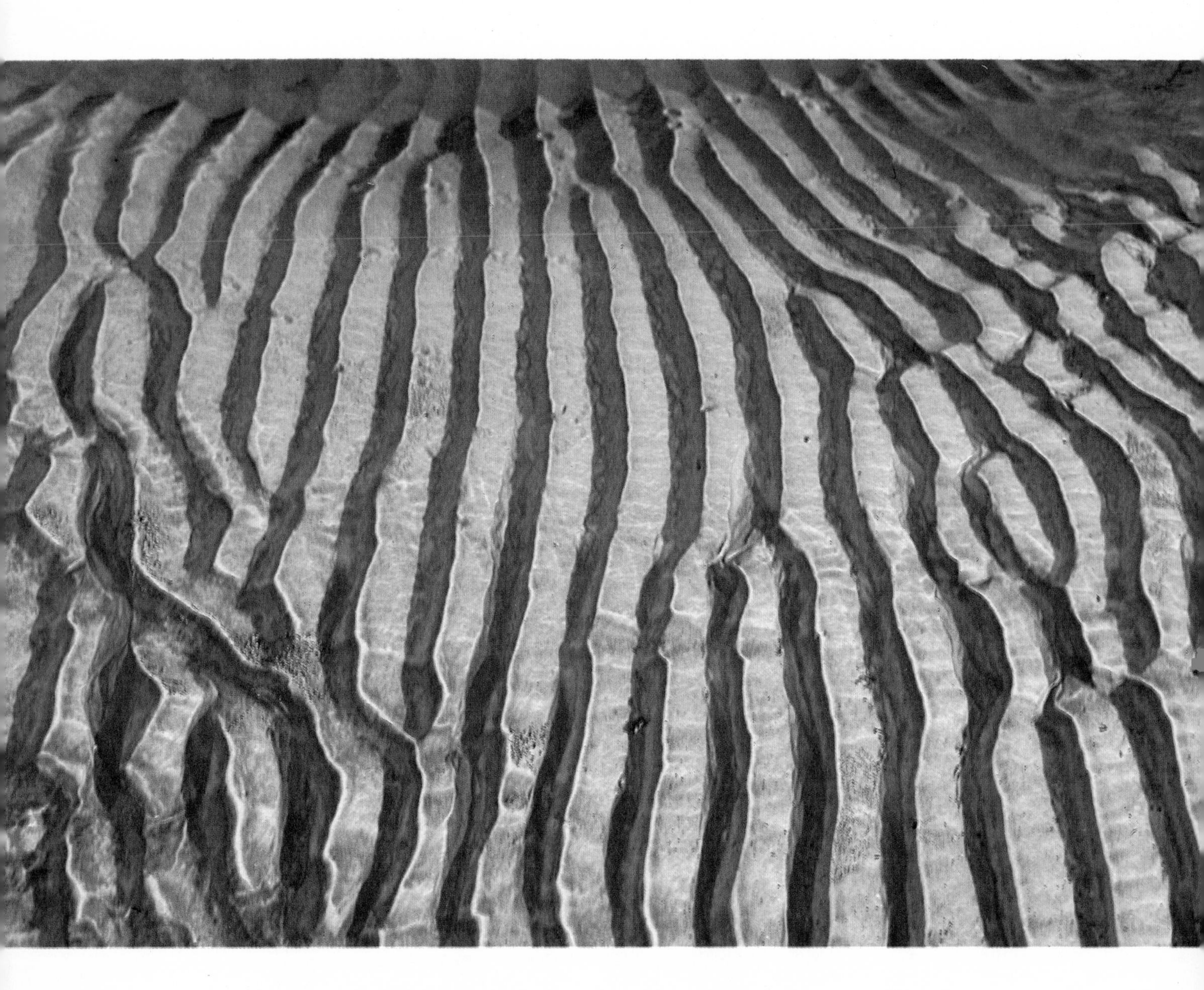

At very low tide the sand is often dark and muddy, and formed into a succession of ridges. A photograph of a small section can reveal a sand sketch, as it were, of rounded cliffs lapped by water. Between the ridges the sea's froth sometimes lingers, looking like soapy water caught in gutters.

As the tide moves farther out, we follow and come upon an area bristling with small tubes, each a miniature chimney leading to the burrow of a sea worm; or maybe the beach is strewn with groups of tiny shells, all firmly attached to a parchment-like tube—in which there is a worm that cleverly protects itself from predators in this manner. We walk in the shallow water. As each wave recedes, the sand bubbles and emits tiny squirts indicating myriad lives below the surface.

Sometimes the frothy water flows around clusters of caviarlike black globules. These formations are made up of the secretions of a type of worm and are known as worm rock.

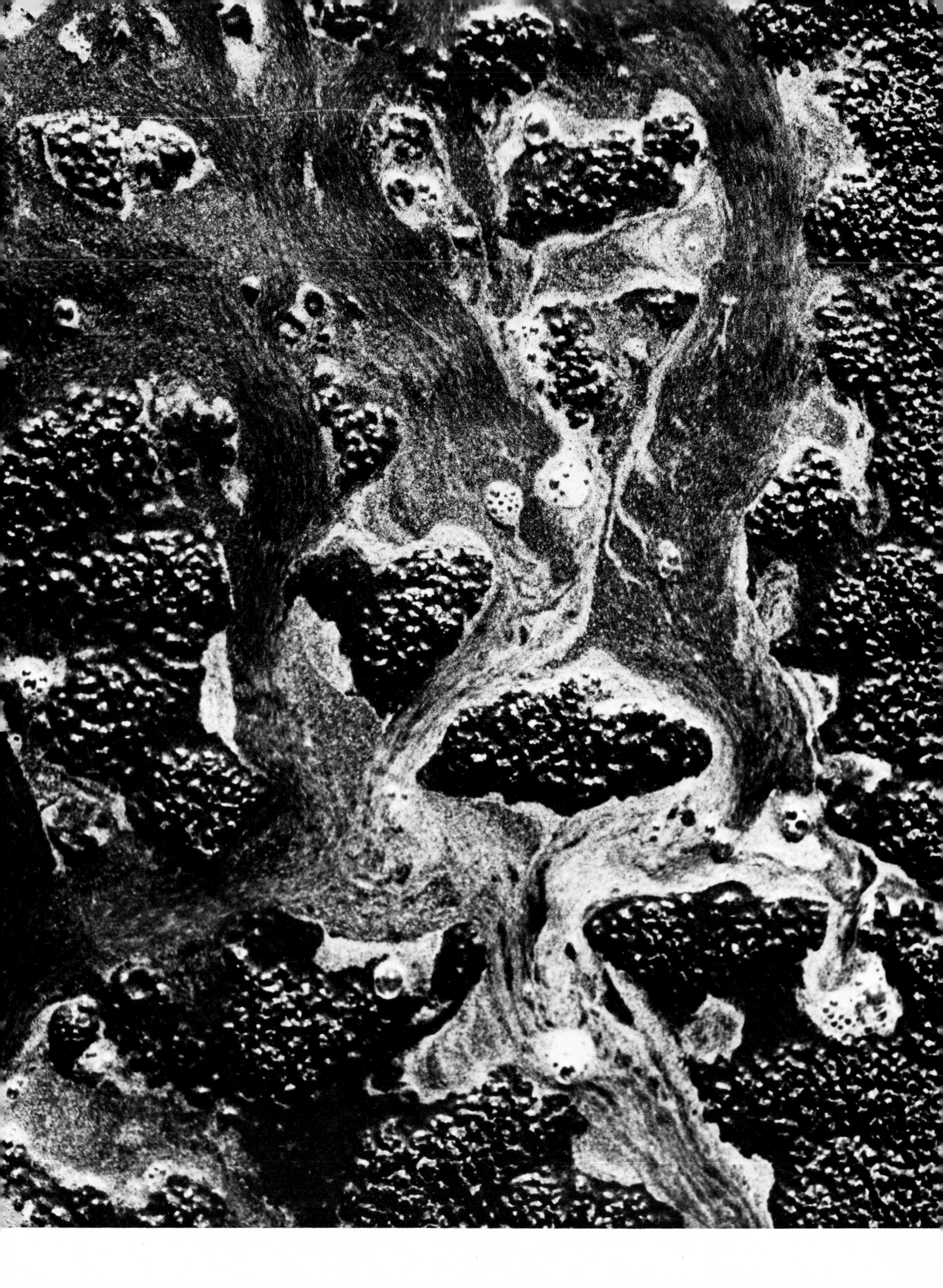

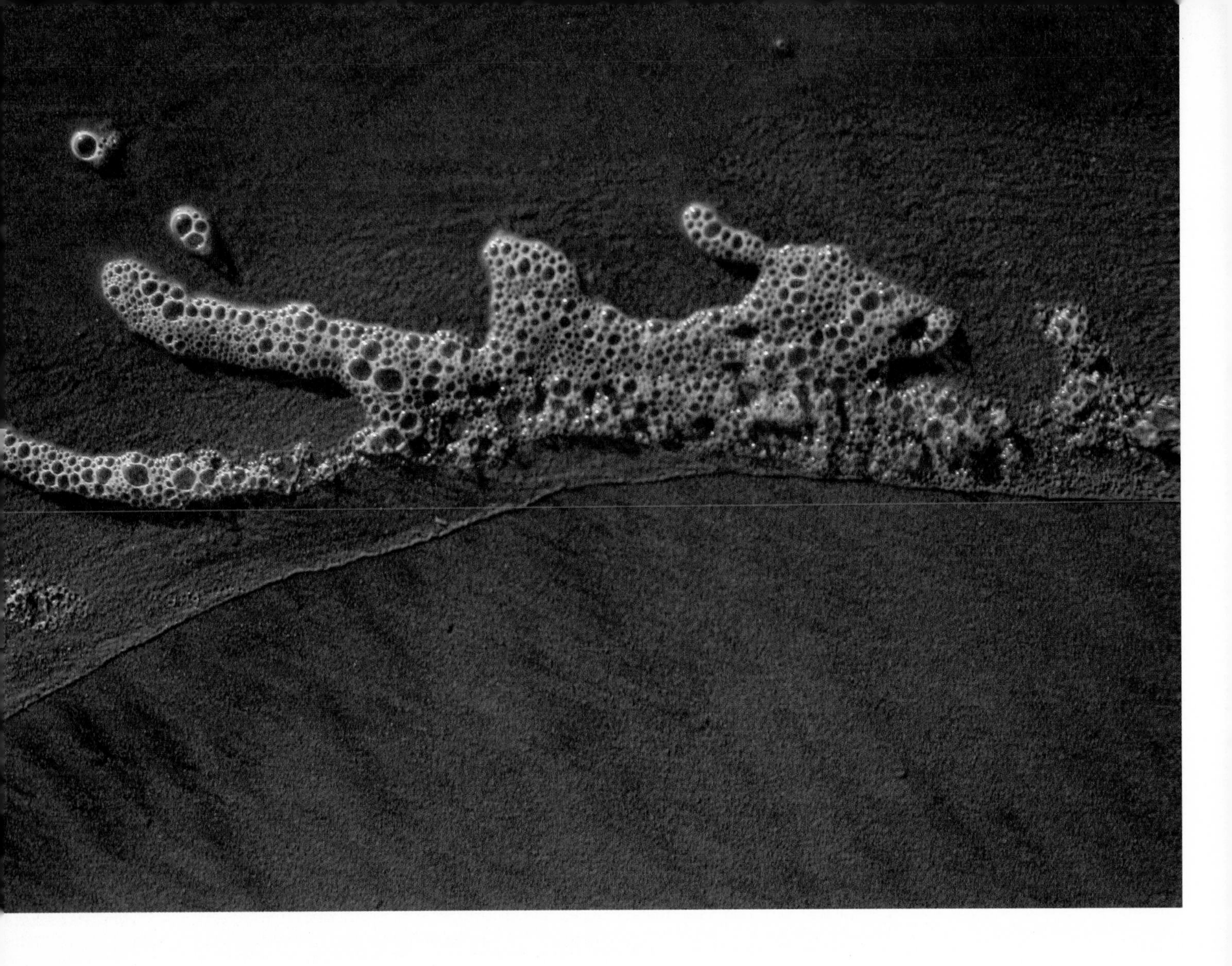

Froth that takes its own shape when abandoned by a wave can be even more engaging, and more transitory; it may be strung out in a wavy but more or less continuous line, or it may take individual shapes to be quickly caught by the camera. In a few moments the bubbles start to break up or are blown away by the wind; the next wave laps in and the froth is gone, only to be replaced by another bubble decoration.

At times, when the sea's currents are lively, foam, water and sand make complex patterns.

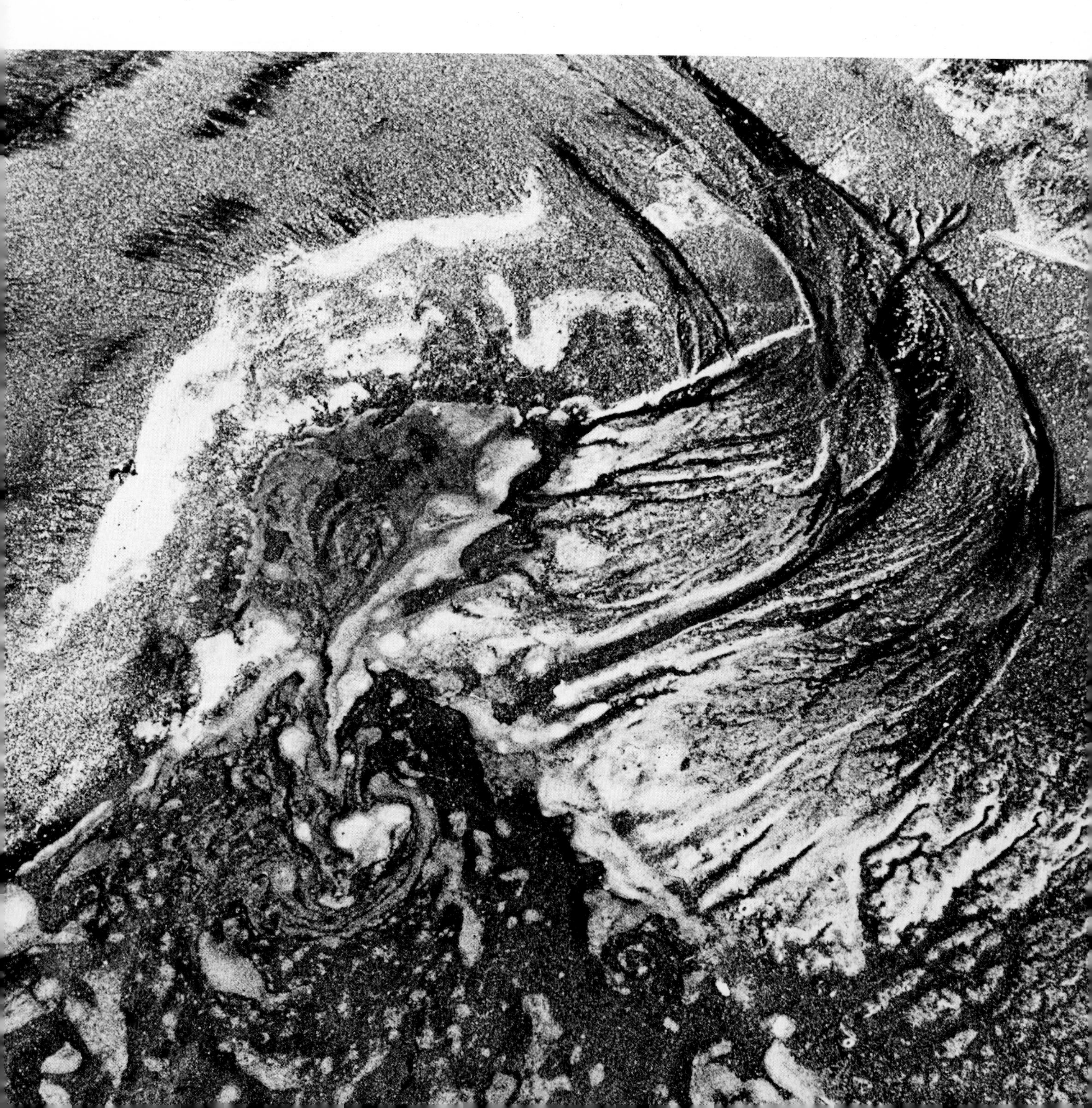

This edge of the sea, the low-tide line, is a fleetingly exposed world. As its time for being part of the land rather than of the sea is short, it is likely to be here that rarely seen things are discovered. Feather stars, sometimes called sea lilies, are such rarities. I have found them only once in shallow, sandy pools at the very extremity of a low spring tide. Without my photographs, I might begin to doubt that I had seen them at all since one authority, Augusta Foote Arnold, says they are never found on the beach. Crinoids, to use their botanical name, are inhabitants of deep water. They grow on stalks, at one end of which is a cup-shaped body with many branching arms. This upper part, at a certain stage, can detach itself from the stalk and the plant state to swim freely as an animal. This was probably the case with those I found, and they were very likely dead, having been washed ashore by some eccentric current.

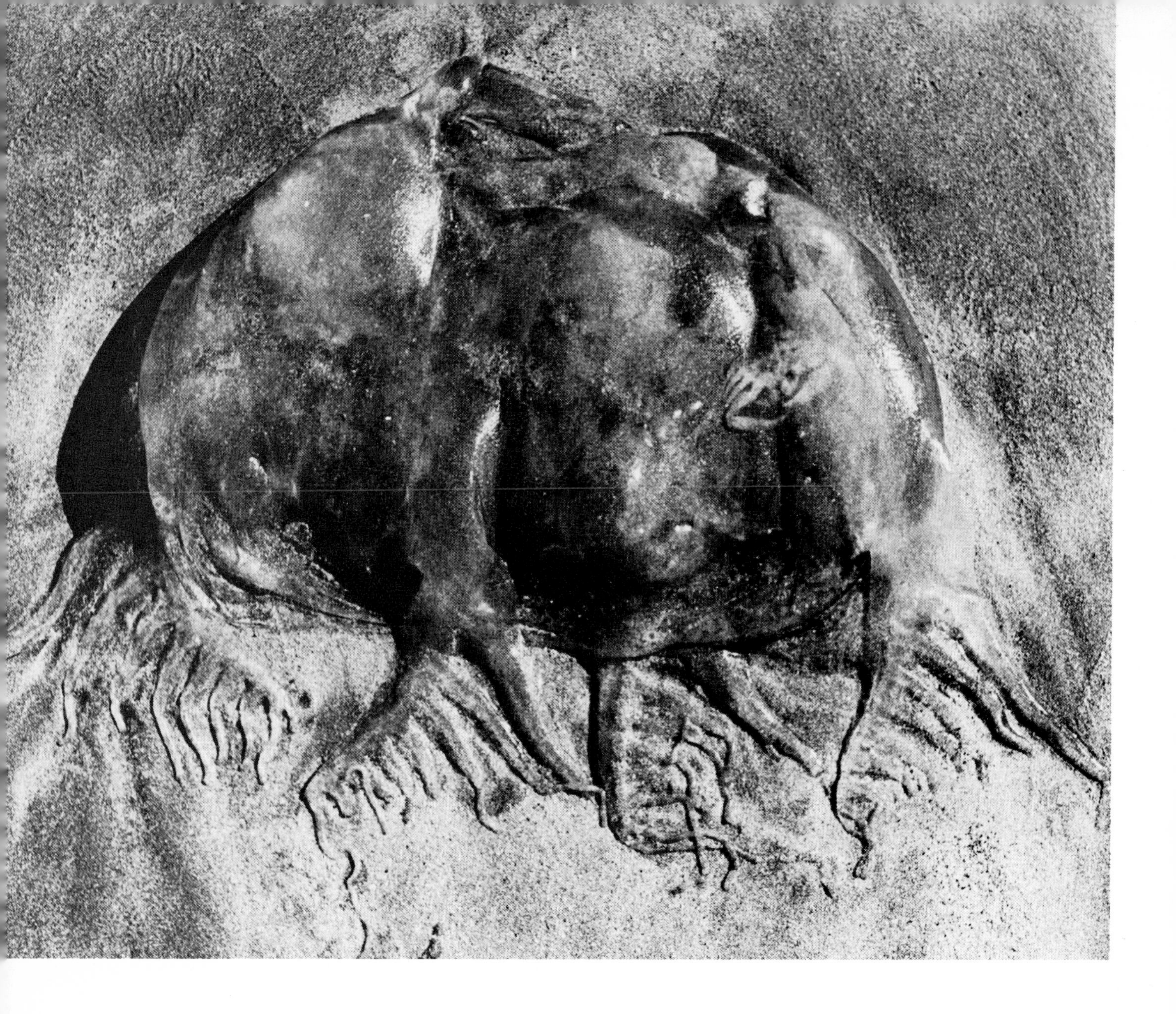

Dead or alive, these feather stars are exquisite, their fringed arms a light golden brown, moving gently with the water. They have existed from early geological times and their history can be traced through successive periods from the imprints of their shapely skeletons on rocks. These fossils are plentiful and known as stone lilies.

With the turn of the seasons, other creatures are thrown up on the shore—great globs of white or gray jellyfish, sometimes two or three feet in diameter, domed like an upturned saucer and unattractive. But even these have an intriguing transparency through which one may glimpse the body of a crab imbedded in their inert jelly fat.

Occasionally the still inflated purplish blue sail of a Portuguese man-of-war may be found on rocks near the surf. The sail, like an elongated balloon, is filled with gas, largely nitrogen, that enables it to float on the surface of the water, its food-hunting, poisonous tentacles dropping below. It has been likened to a fishing boat with a drift net of high-voltage wires, so deadly is its sting to almost all fish and sometimes to humans. Biologically it is a curiosity. It is thought to be not one animal but a colony of individuals, none of which could survive independently. It is most certainly a customer to be avoided when bathing—better to find it with float deflated though still iridescent, its bright blue streamers disheveled on the sand. Less dangerous are the small circular discs fringed with blue feelers like blue-rayed suns. A kind of jellyfish, called Porpita, they are found caught in pools or strewn along the tide lines a little before the monsoon, along with a few starfish bleached white by the sun. All of these, if they are up on the dry sand, are dead or nearly dead, for their last act in life is to allow themselves to be cast up by the sea.

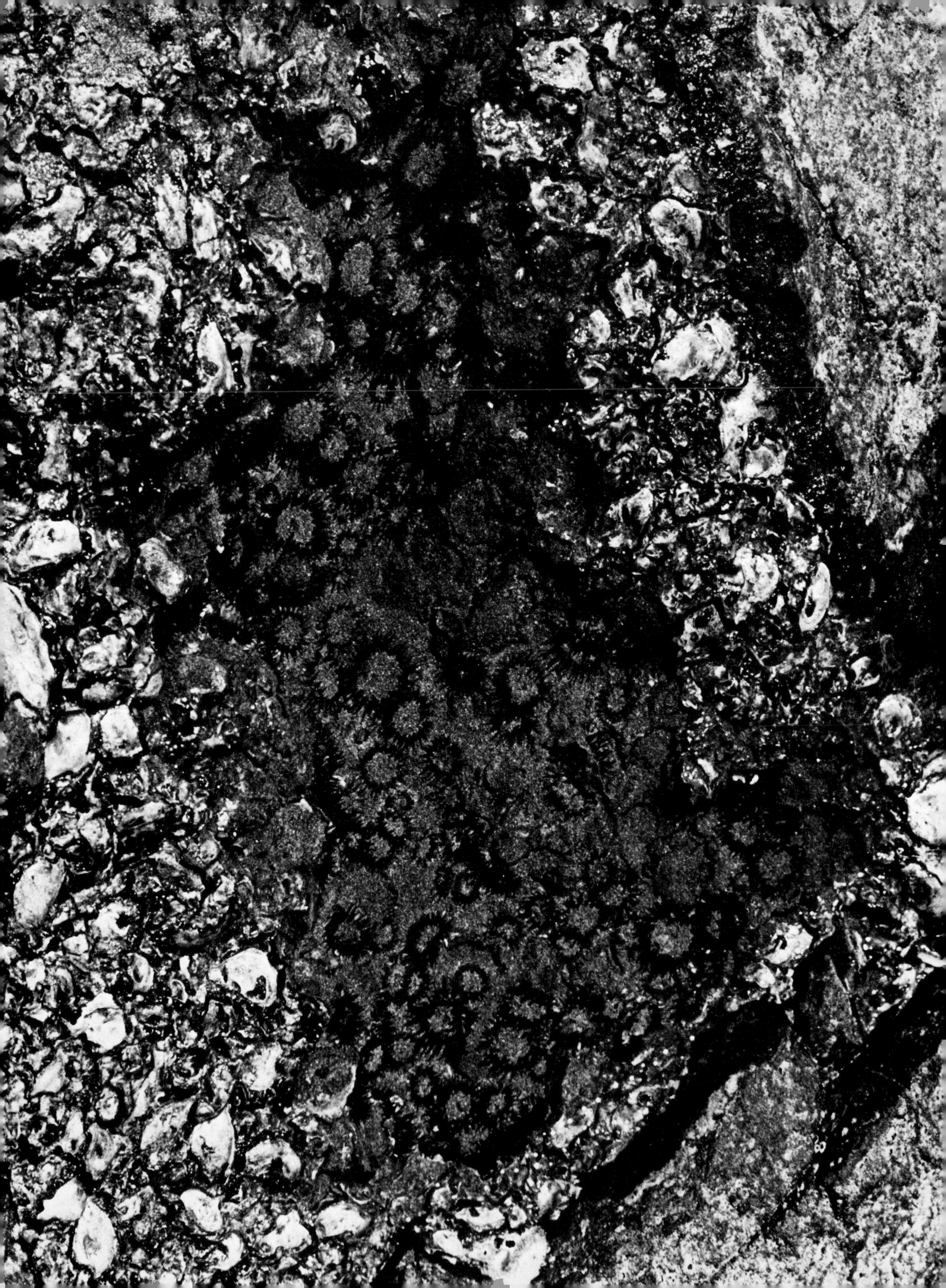

Then, too, there are the less willing cast-ups that come in unwanted in the fishermen's nets. Among an assortment of inedible fish there may be red-brown stingrays, and black-and-white-banded sea snakes, always lying in graceful curves and harmlessly dead.

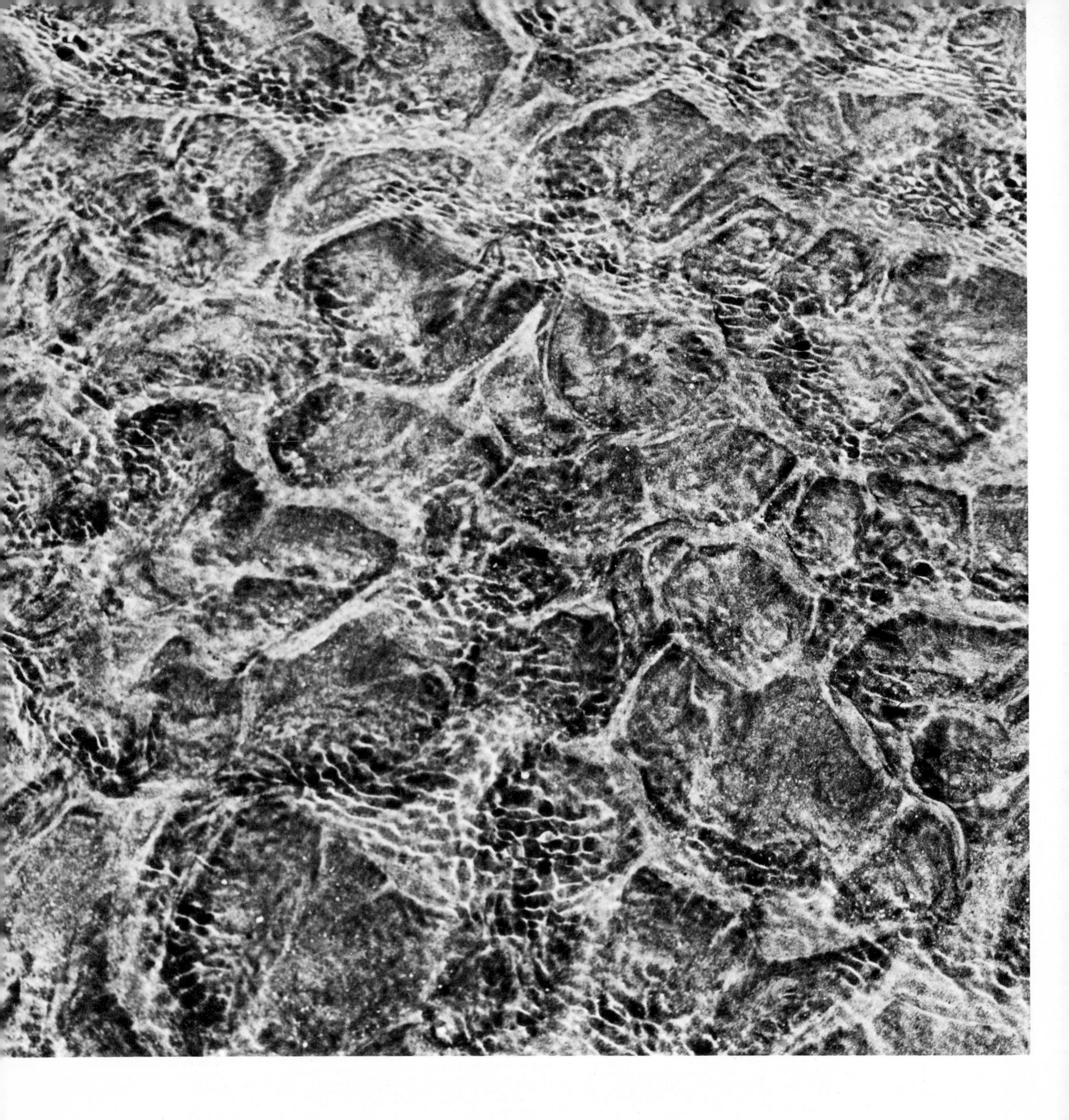

After the tide has ebbed, the water still runs in certain gullies, and should the current be fast it is as if the water piles up on itself in a series of bumps and ridges all sparkling giddily in the sunshine. Looking directly down on such fast-moving water we sometimes see a pattern from another realm—that of science and the microscope—a photomicrograph of some section of a plant.

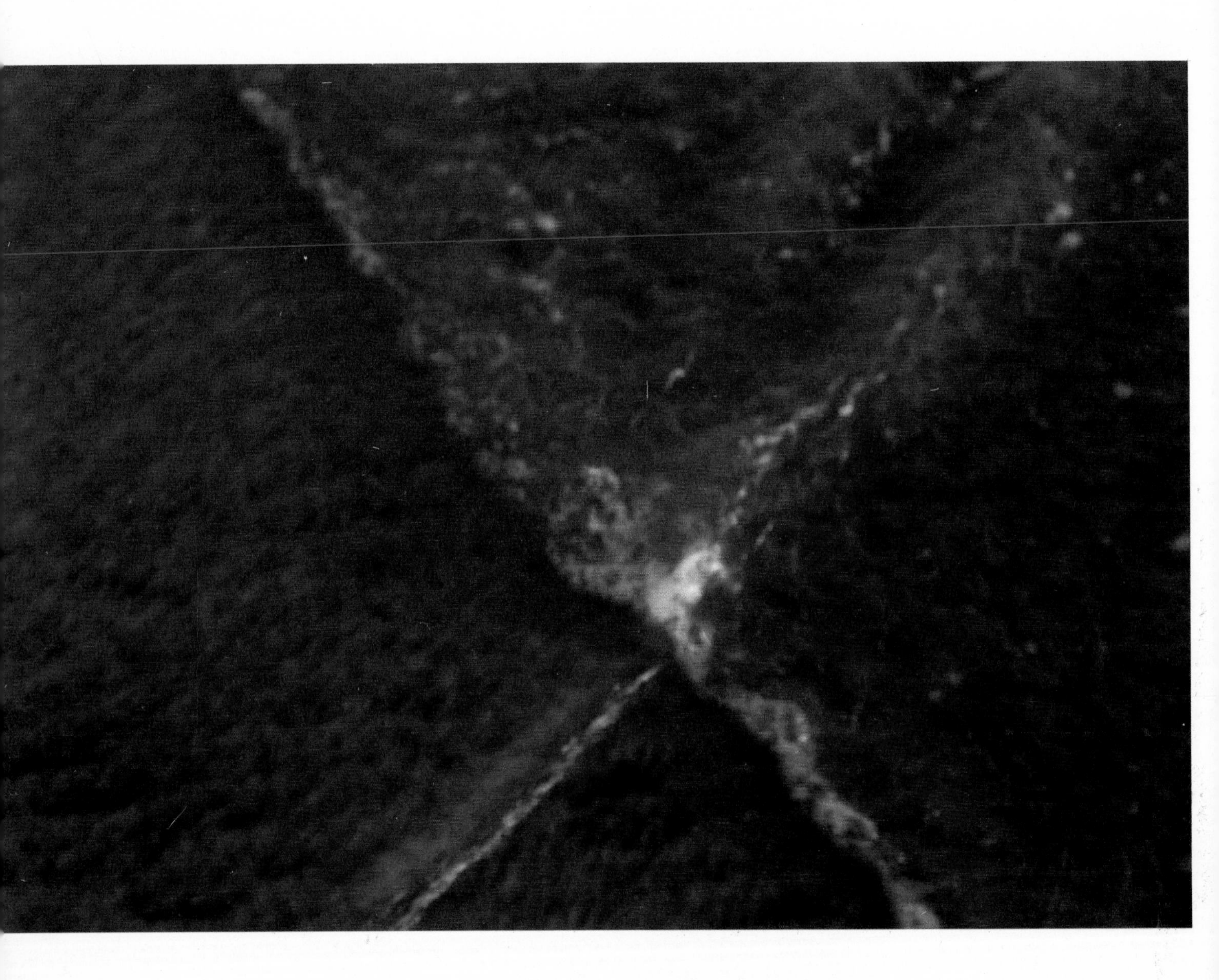

There is movement on the shore apart from that of living things—in the flow of the waves themselves and their crosscurrents that collide and run at right angles, each keeping to their own direction.

Further kinetic effects are produced when the wind and water play with different shades of sand, crossing them into diamond shapes, blending, streaking, as if some movement had just passed hurriedly by.

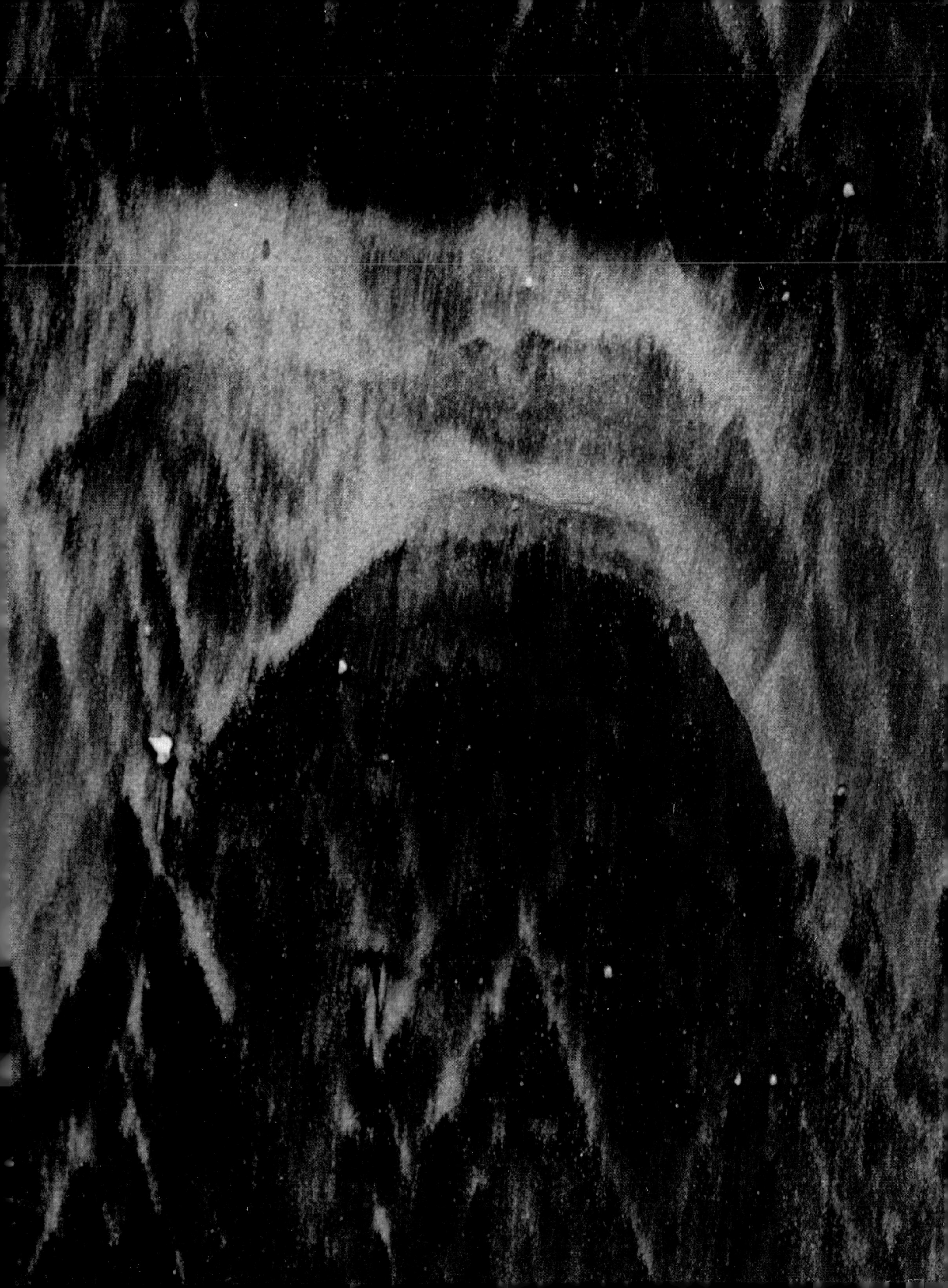

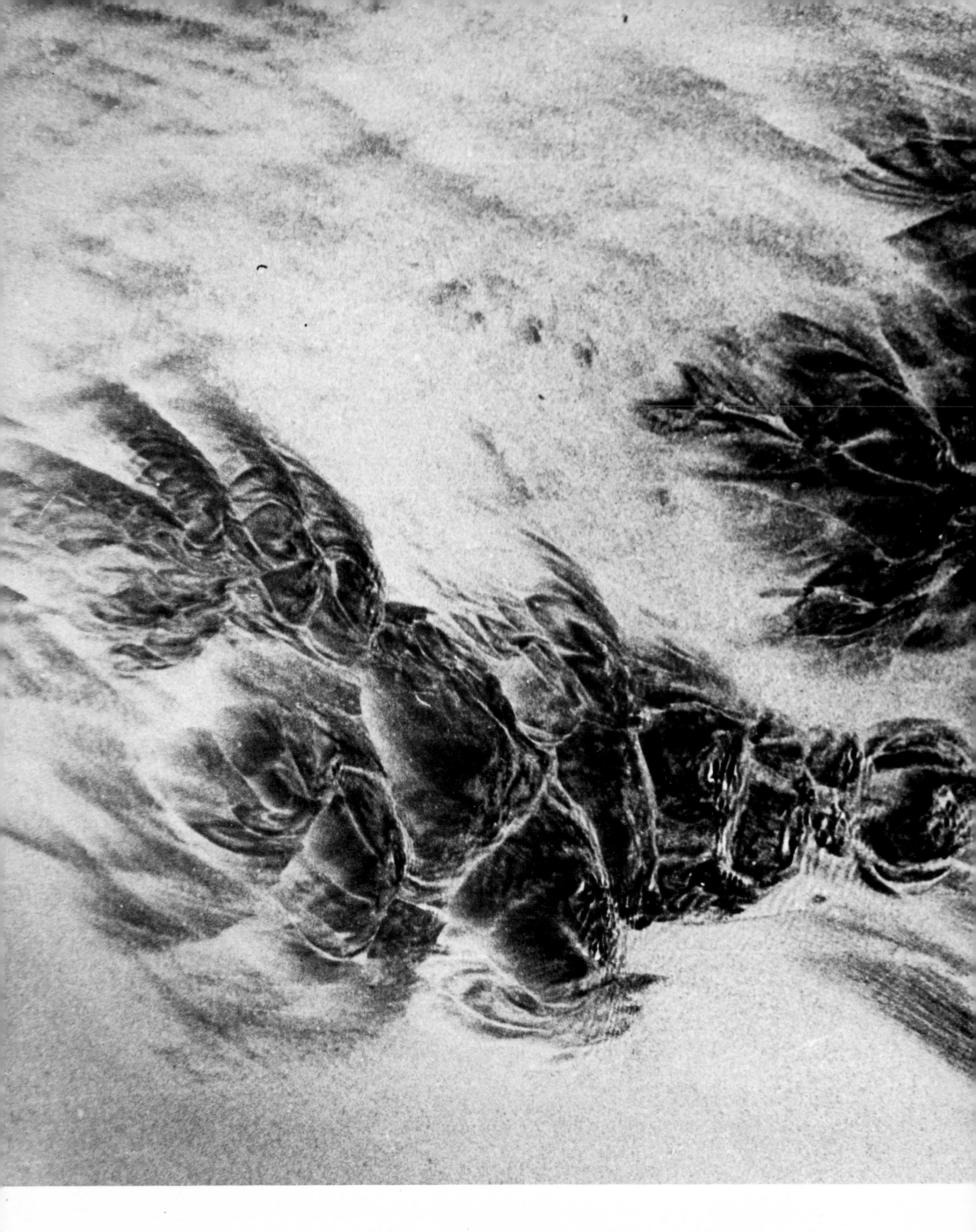

When the beach stream moves a little slower, the sand of its bed is drawn into flamelike designs, bird wings, plumed tails; all is in motion and somewhat blurred because the shapes of the water ripples are superimposed on

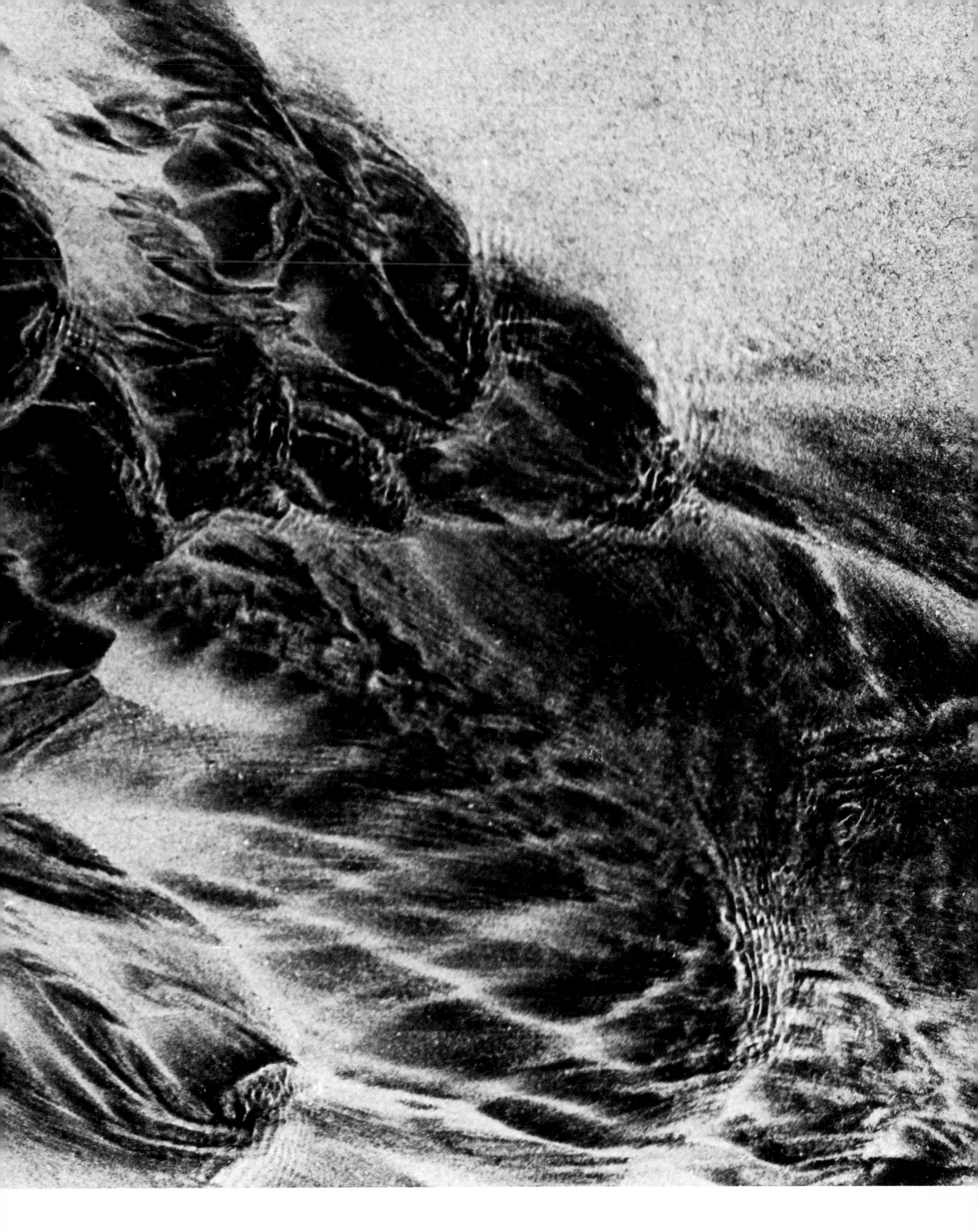

the sand patterns beneath. It is wonderfully alive like something windblown, something turbulent, but another photographer could not be blamed if he were to think the camera had been shaken.

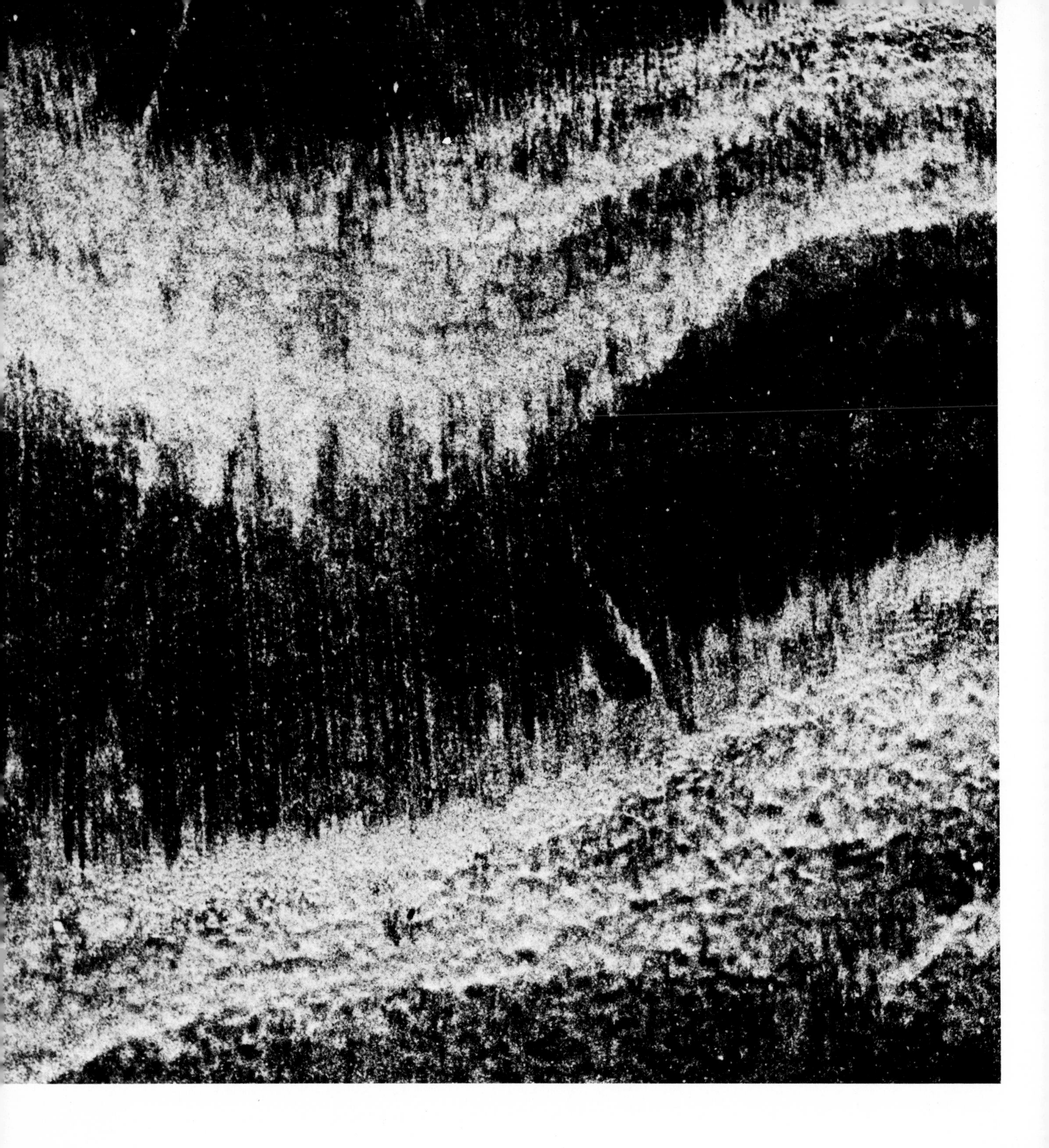

On some of these stretches of sand we may notice a greenness not usually associated with beaches. It is caused by a deposit called glauconite which occasionally produces a green film that lies across the sand like a diaphanous watercolor wash.

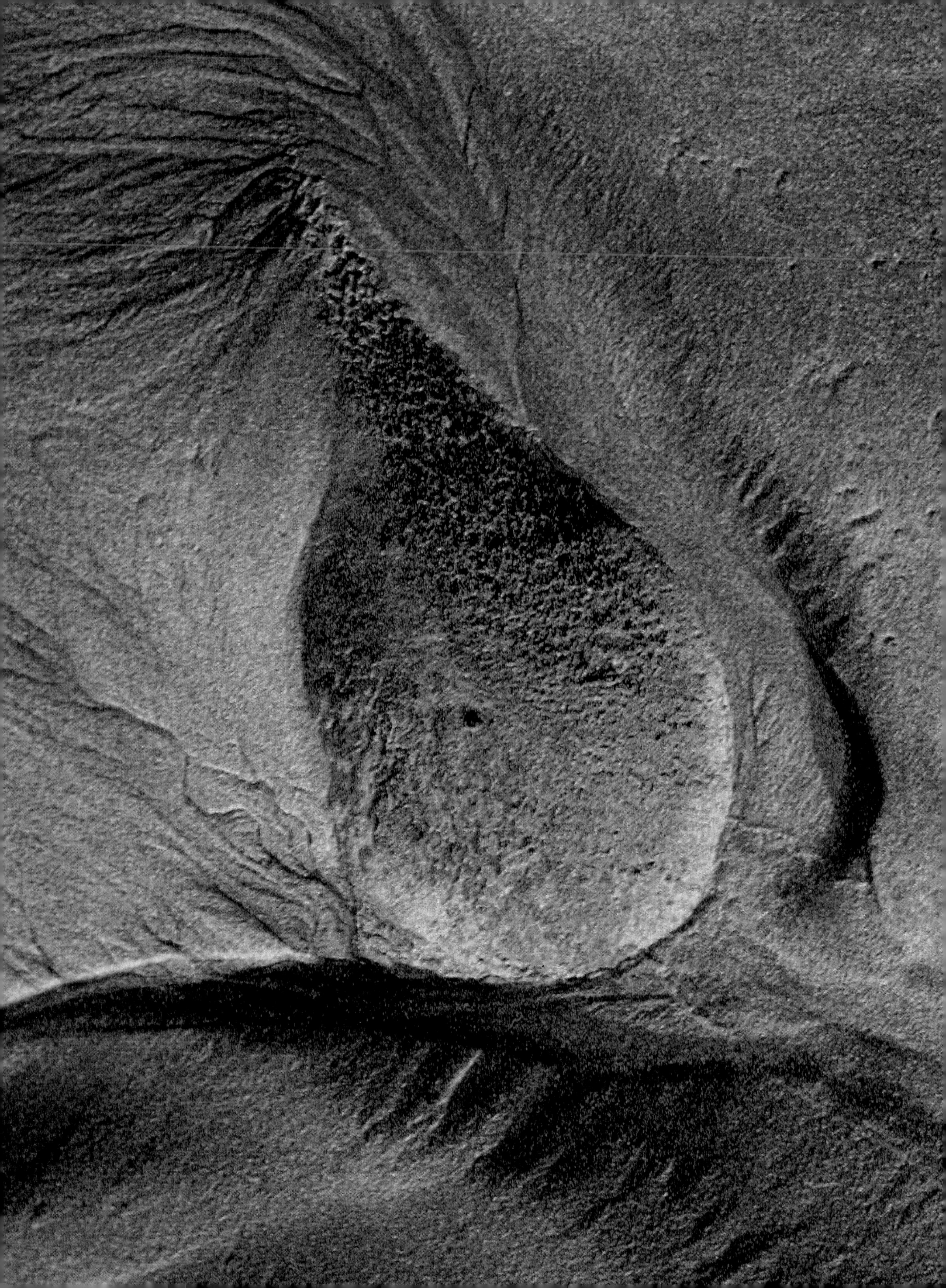

A rock, a piece of driftwood, a pebble, the smallest shell, all can slightly alter the face of a beach. Around even a modest rock, as the tide ebbs, the water will swirl and make a hollow, so that for a time a lake surrounds the island rock. As the water drains, inevitably patterns form, dark streaks radiating into the hollow or one long causeway heading from it toward the ocean.

Here in one small area are three of the beach's chief designs—the soft draping of sand in the center flanked by treelike forms and at the top crisscross diamond shapes.

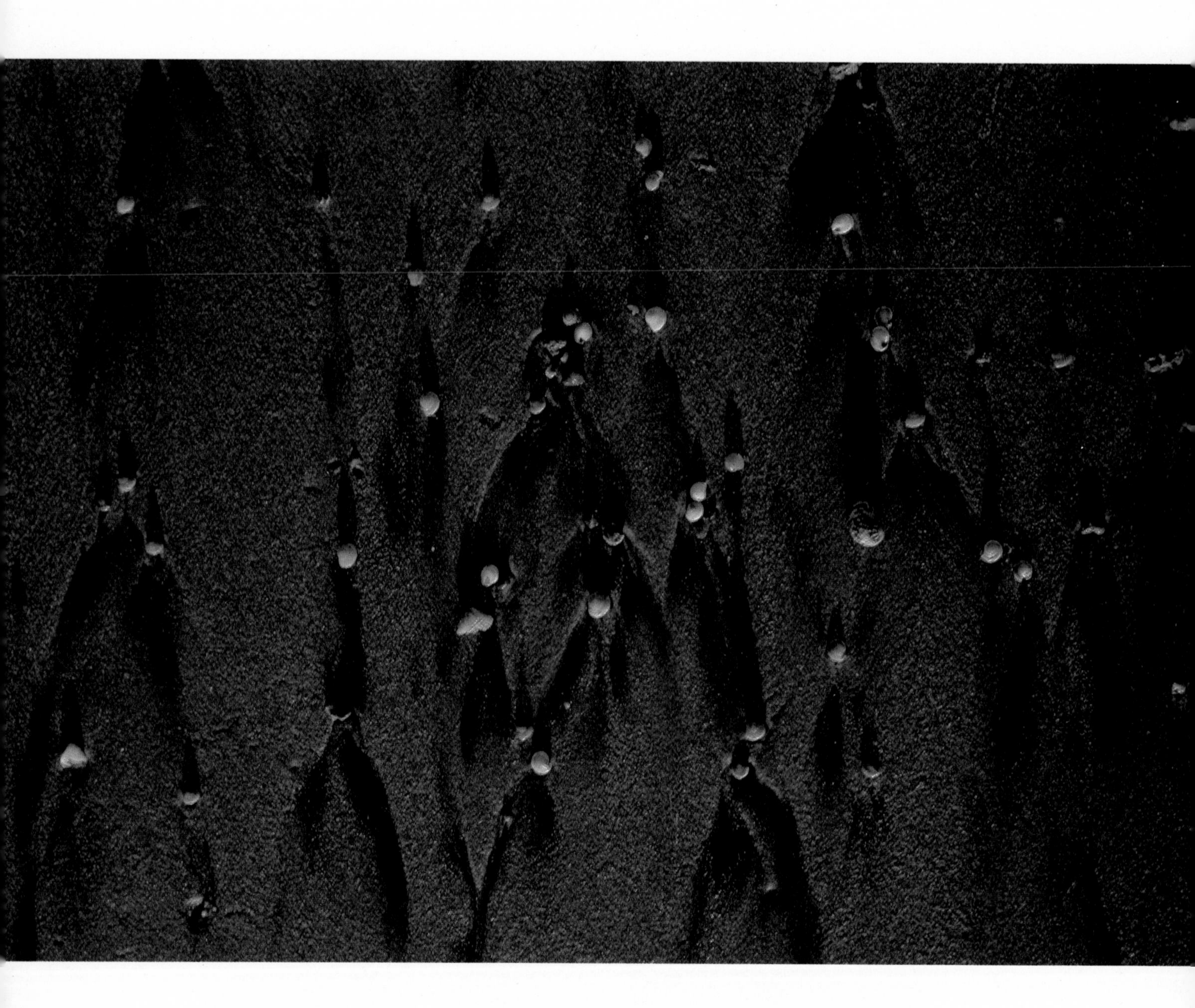

Lighter obstructions to the water, such as a sprinkling of pebbles and shells, usually have dark marks darting from each side but directed seaward. Like beads on cloth, the shells, often bright pink and jewel-like, the pebbles white or orange, glow on the warm, sunset sand trailing their double "shadows."

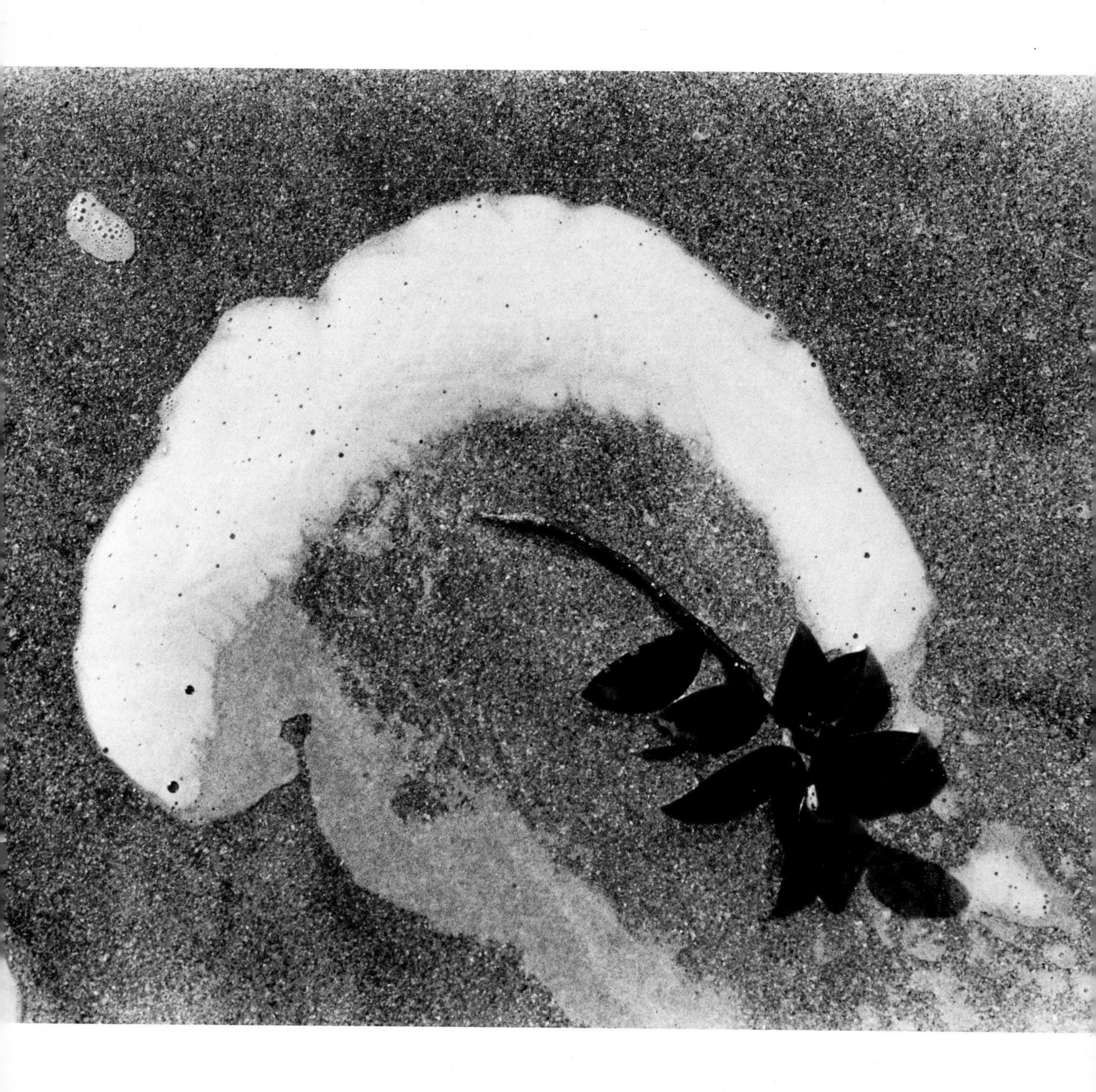

Discarded flowers, lost leaves, garbage, all are pattern makers, the dark sand at times slipping away in curves like a wishbone or drifting into the shape of a tapering ghostly figure.

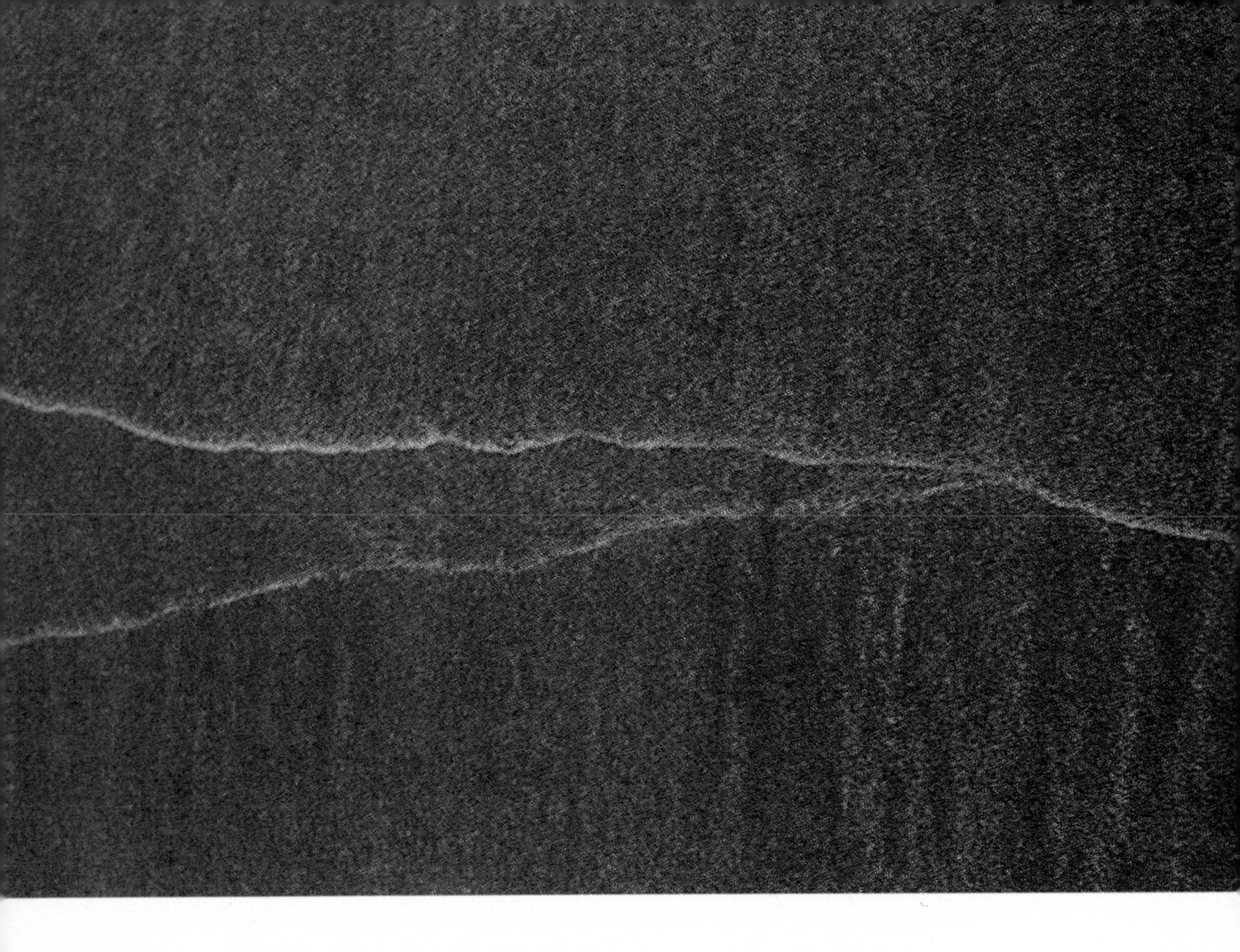

To be first on an untrodden strip of sand gives a sense of elation, almost of awe. This delectable smoothness, marked by nothing but the faintest of lines made by the waves, should perhaps remain unfootmarked. But it is as irresistible as is the greater unknown to any explorer. There are virgin beaches just as there are virgin forests. The world explorer may find an unknown tribe; we, a piece of amber, a shell unknown to us at least. Sometimes far down the beach or near an estuary the sand is soft. Because we are the first we do not know how much softer it may become. Could there be real quicksands? Today running feet sink in only about three inches. We turn and look. The footprints have already spread and become vague holes. Is this then how Yeti footsteps became so intriguingly big in the high Himalayan snows?

Where the sand is firmer but still wet, toe prints dribble what look like branches. Higher up the sand is only damp, and a sharp, well-formed print will remain until smothered by another or until the sea's return.

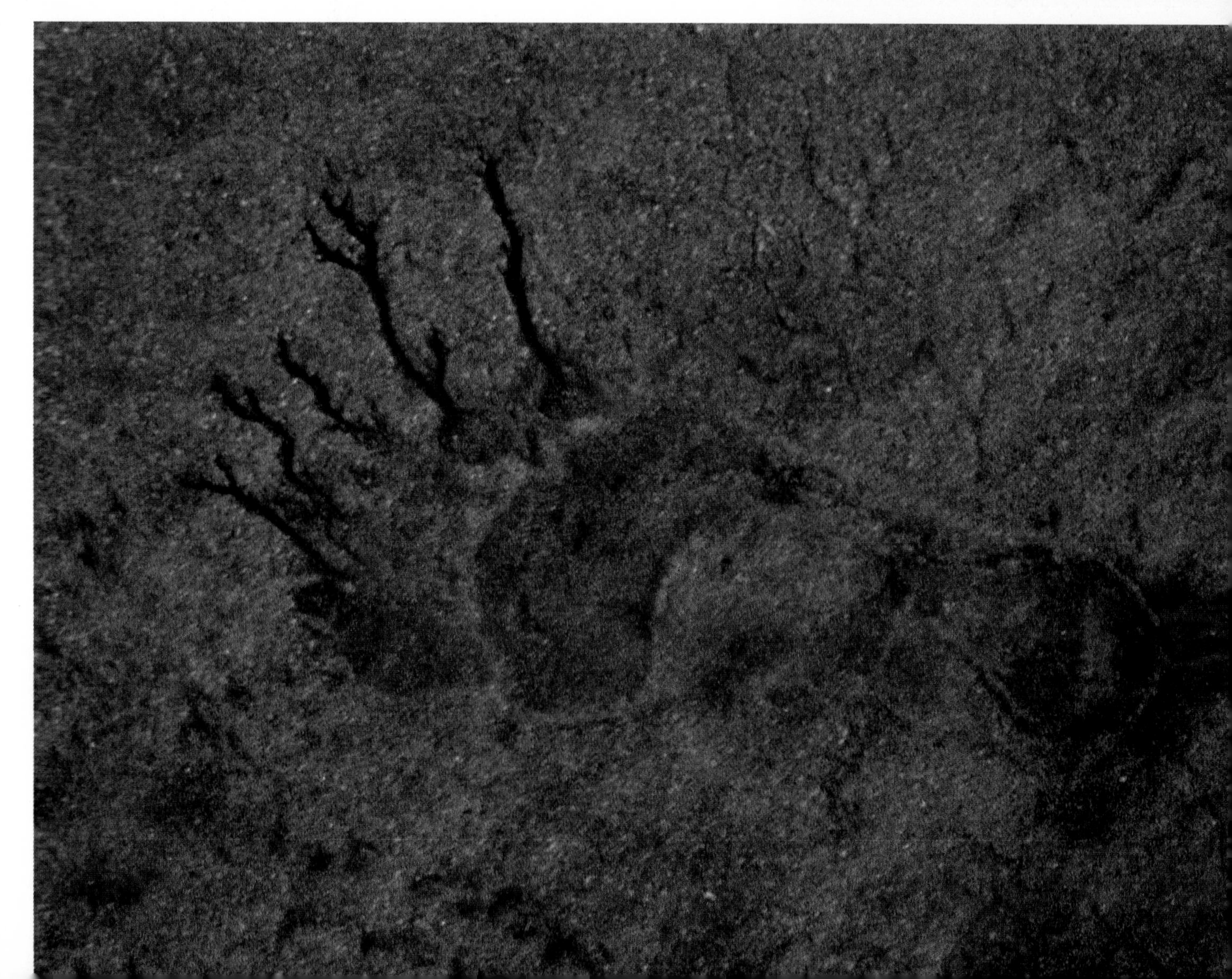

In the vicinity of estuaries there may be stagnant pools of water caught in a hollow when the river changes direction, or caught because the waves build up a sandbank. Sooner or later the pool may disappear either through drainage or because the river or the sea alters the lie of the land once more, but in the interim, algae are likely to form in the still water. Within the term algae, all seaweeds are included, as well as the very simple unicellular growth that throws a green scum over ponds whether fresh or brackish. These simplest of all plants are known as thallophytes, meaning that they have no flowers, leaves, or stems. Each individual is a single microscopic cell that divides and subdivides and so spreads itself like a green mat. These kinds of algae are usually as smooth as the water they cover, but now and then some are mottled with white and somewhat bumpy. All may be stirred by the wind, an animal, or a stick into a pattern.

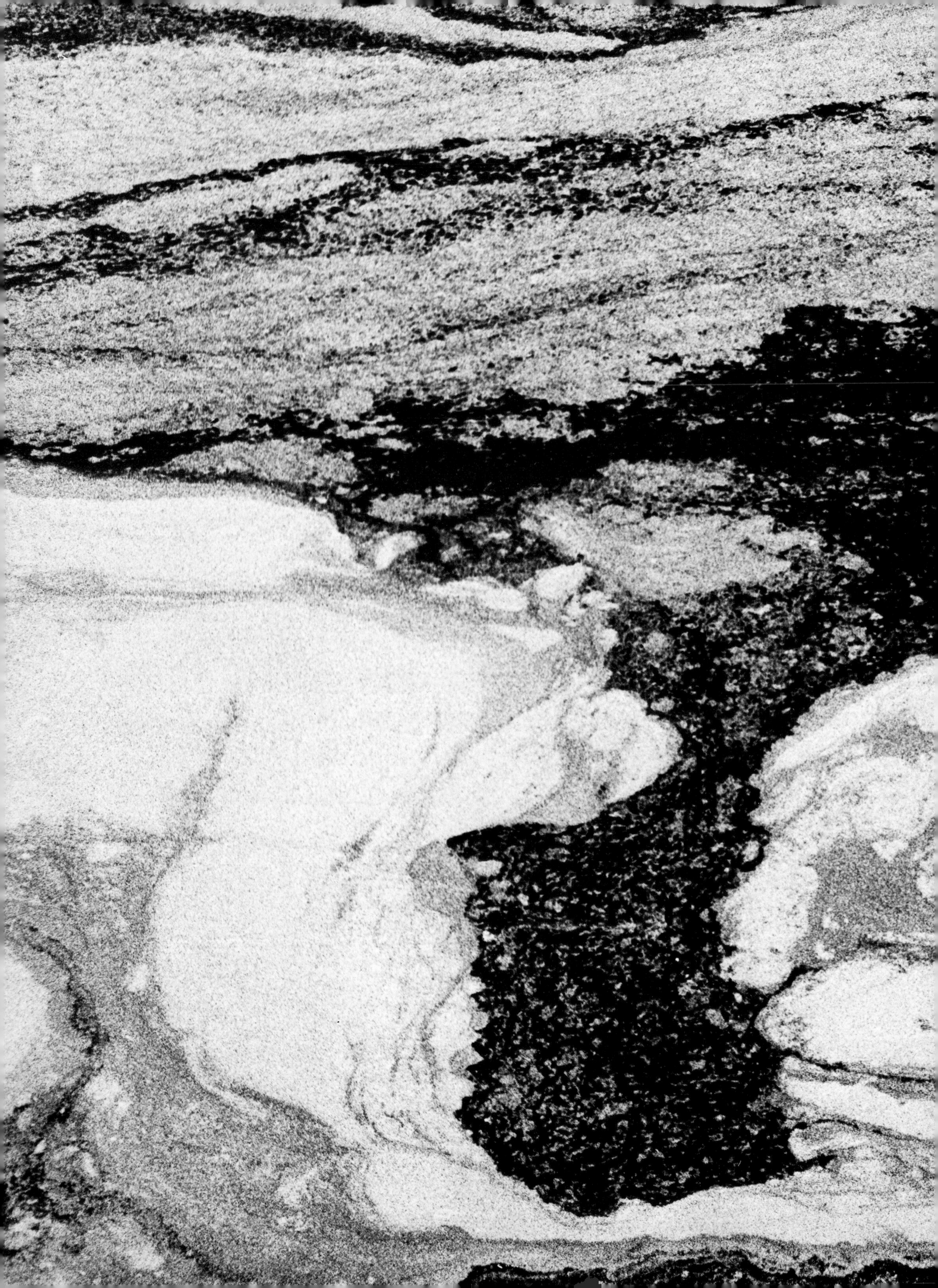

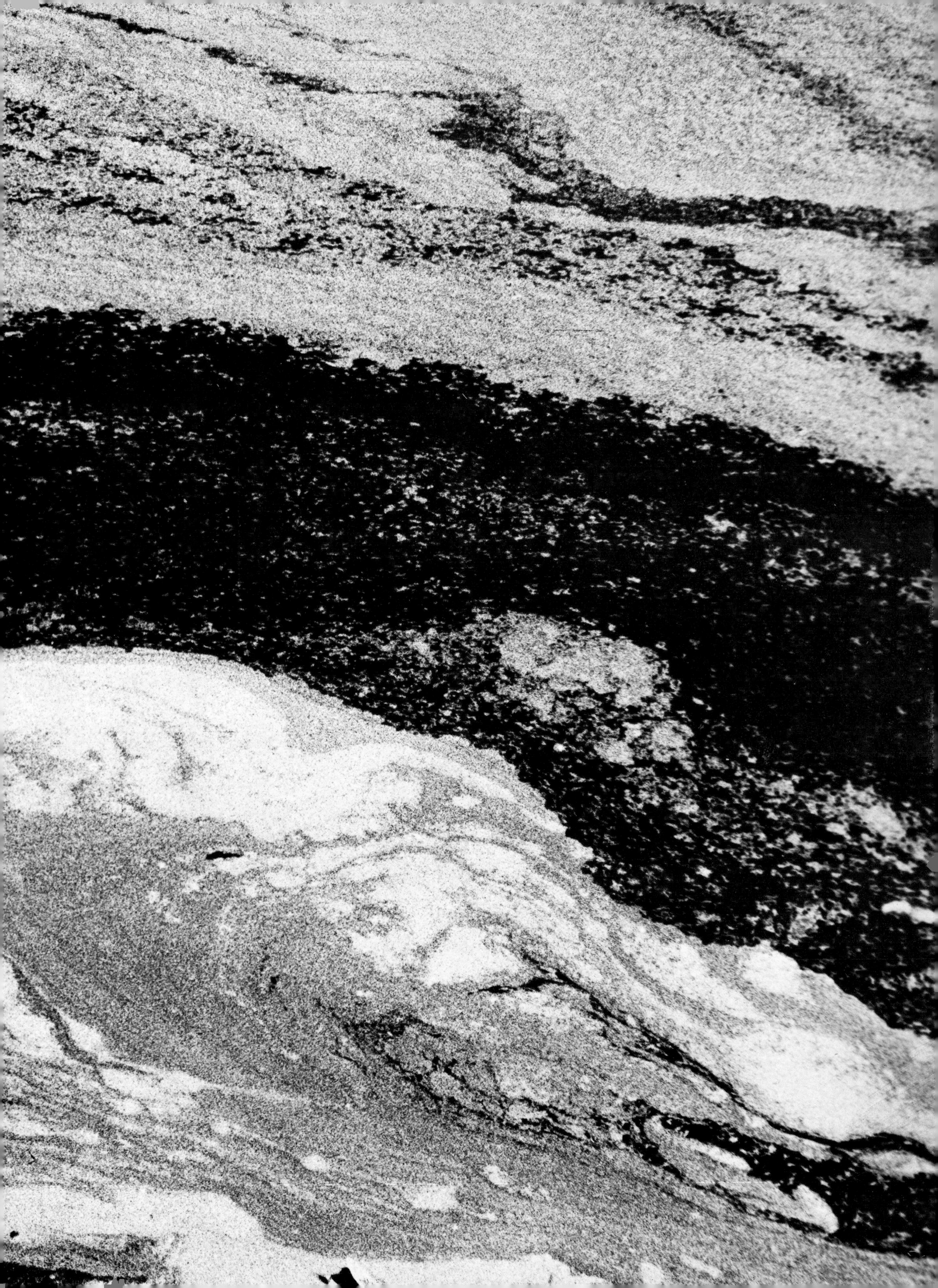

Wind causes the sea to surge and leave a hundred pools; the dazzling stripes diminish or vanish only to reappear a few days later. Energetic waves forge a gully that turns and becomes a long inland lake parallel to the ocean, and this lake will have its own set of scalloped beaches.

Hardly a day goes by without these beaches taking on a new look. They become rumpled and banked where before they were smooth.

The creeks at either end of the strands change their geography frequently; sandbanks are pushed up, islands and sandy headlands appear; dwarf precipices are built and then undermined by the persistent caress of the water.

Then patterns may form on the banks in curious rows and arcs: weird monsters gathered in a dance, the sand in this case looking permanent and hard, almost suggesting the fossils of long-disappeared plants.

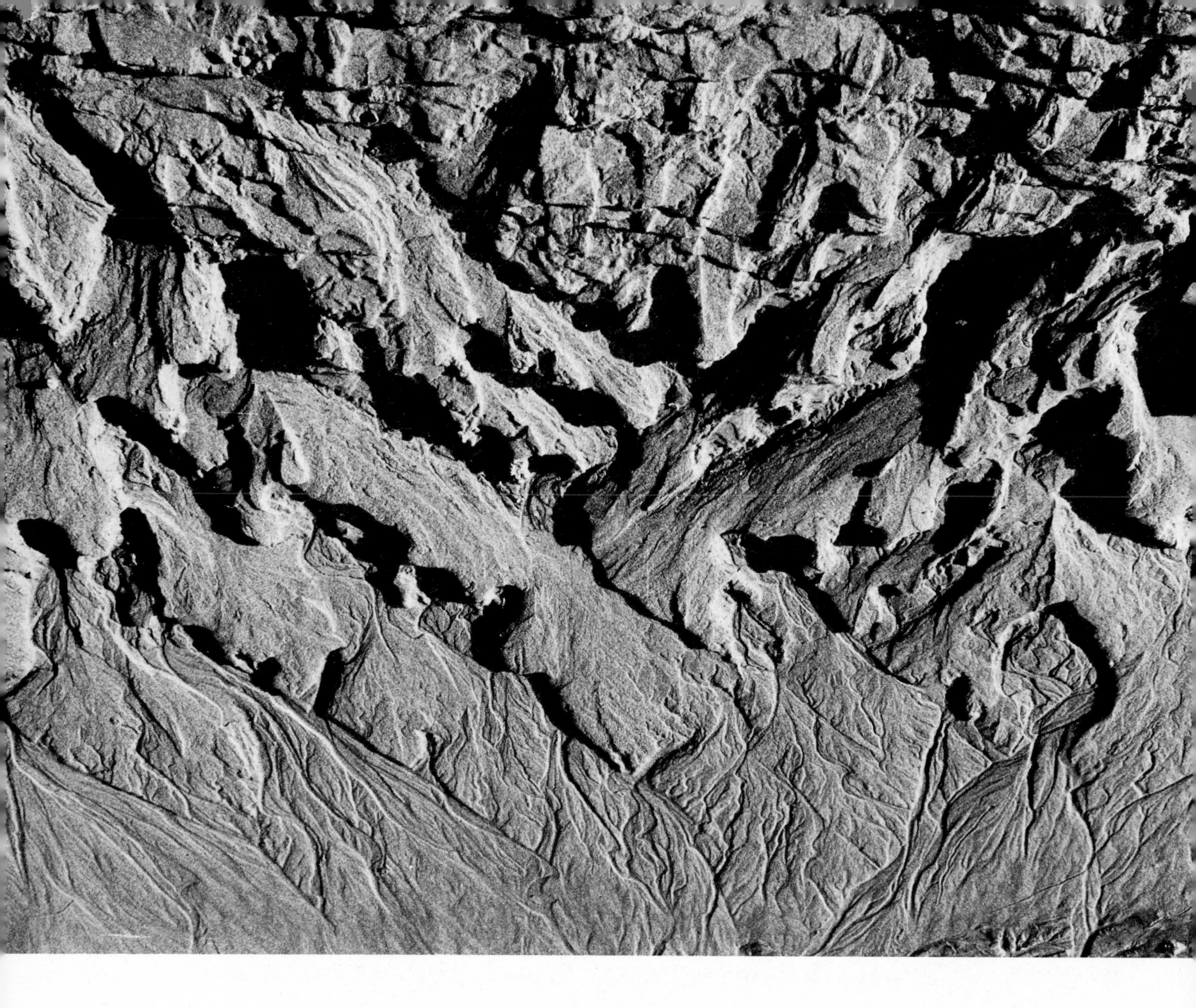

Sand, given a little dampness to hold the grains together, can resemble a rock cliff studded with caves or the weathered vistas of Death Valley.

The laws governing wind, sand, and water are fairly constant. Any desert landscape or one involving estuaries seen from the air is here on the beach in miniature, reproduced to perfection, the small imitating the large with fascinating exactitude.

And, as has been already suggested, sand patterns can resemble quite different aspects of nature such as plant growth, trees looming through mist, the movement of hair.

There are other similarities: tree bark, the grain in cut and polished wood or in marble, even the muscles of a human leg in an anatomical drawing, or the kinetic look of certain handwoven materials.

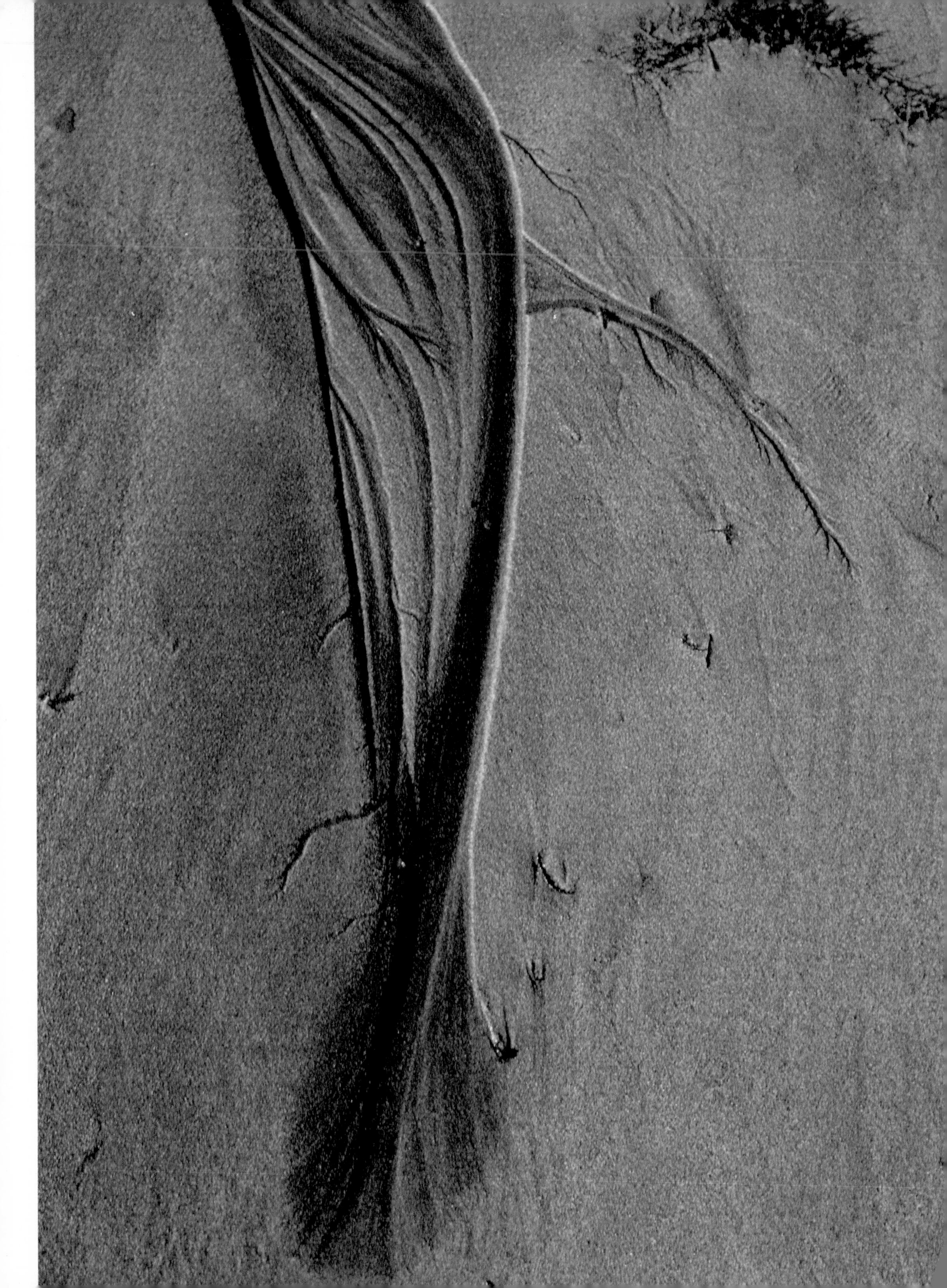

It can be said then that on the beach, as in clouds or in rocks, patterns and images are there and resemble, in our imaginations at least, a wide variety of things; that the delicate formations in sand, so easily overlooked at our feet, really do match those on a larger scale seen from an airplane and perhaps from a rocket; that throughout nature there are patterns and that we, too, make them in the many things we do; that animals, insects, and birds make them, and that modern science has shown us a profusion of these natural arrangements, never before seen, from the microscopic to the infinitely distant. From these newly discovered patterns surely comes further confirmation of the basic similitude of substance.

Behind, beside, and through all this additional knowledge is the interest taken by scientists in the shape and relation of things to one another, of the smallest with the greatest, and between elements seemingly dissimilar. And it is not only scientists who are probing these mysteries but philosophers, artists, and mystics as well. Indeed, it is likely that mystics were the first probers.

Adding to the magic of the shore is the fact that it was the primeval scene of the first beginnings of life on this planet. Some organism, drifting in the shallow water, went through a change, reproduced itself and yielded, in Rachel Carson's words, "that endlessly varied stream of living things that has surged through time and space to occupy the earth." Here too, at this meeting place of land and water, we can observe those provocative likenesses to other substances etched in the sand by the water. Our imaginations can move in both directions, into space or into our own cells. The beach lover, particularly the student of patterns, may perhaps be excused and even understood if he thinks that the sand sends messages and that intertidal spaces are windows through which we see replicas of things vast and very far and of those too minute to be seen by the unaided eye.

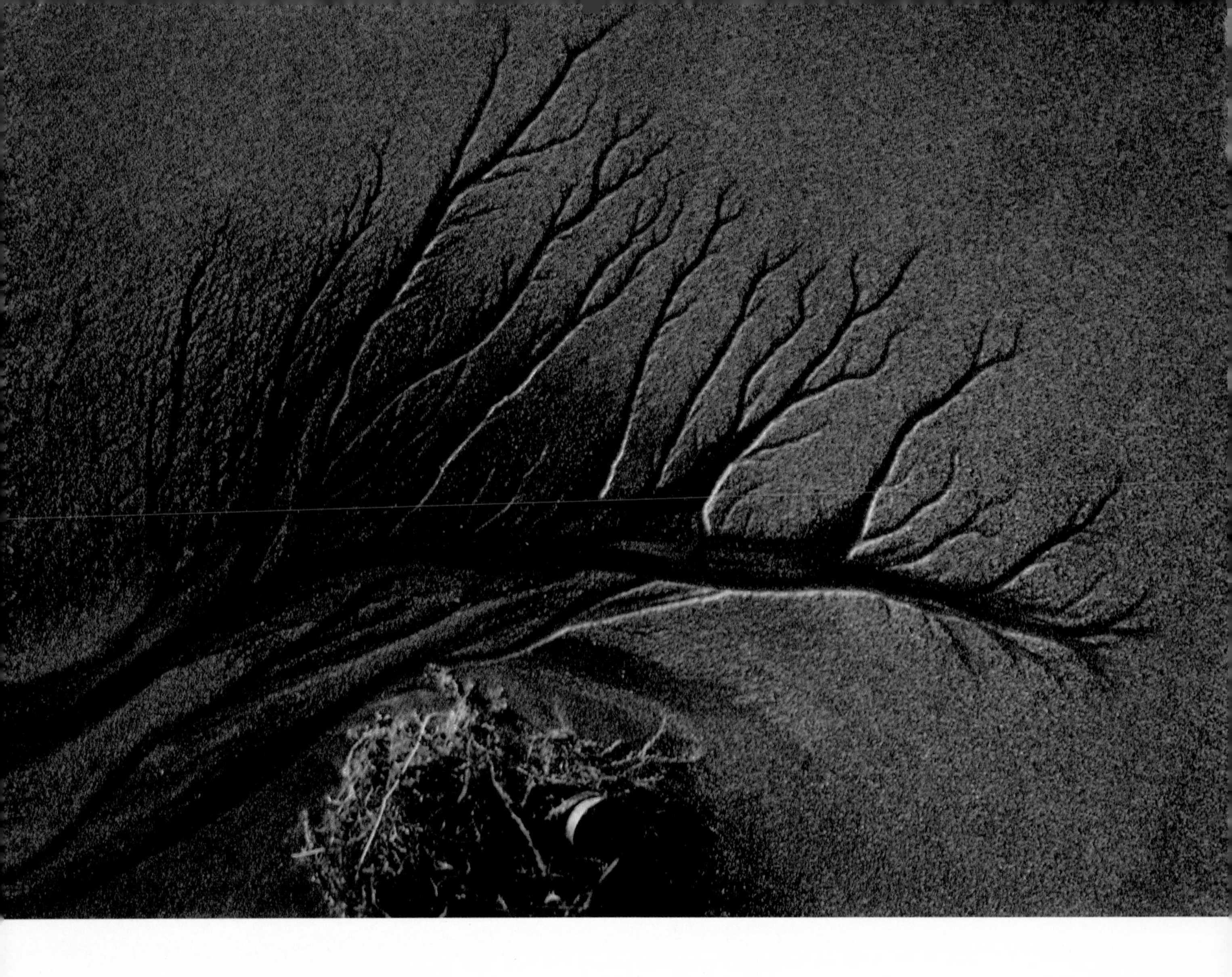

The world is a whole, a mass of amazing similarities that are at once continuous and transitory as is all life. Possibly these musings are unconsciously present when we simply open our eyes and take delight in patterns made by small crabs or trickles of water on sand.